Bent Functions

Bent Functions
Results and Applications to Cryptography

by

Natalia Tokareva
Sobolev Institute of Mathematics
Novosibirsk State University
Novosibirsk, Russia

AMSTERDAM • BOSTON • HEIDELBERG • LONDON
NEW YORK • OXFORD • PARIS • SAN DIEGO
SAN FRANCISCO • SINGAPORE • SYDNEY • TOKYO
Academic Press is an imprint of Elsevier

Academic Press is an imprint of Elsevier
125 London Wall, London, EC2Y 5AS, UK
525 B Street, Suite 1800, San Diego, CA 92101-4495, USA
225 Wyman Street, Waltham, MA 02451, USA
The Boulevard, Langford Lane, Kidlington, Oxford OX5 1GB, UK

Notices
Knowledge and best practice in this field are constantly changing. As new research and
experience broaden our understanding, changes in research methods, professional practices,
or medical treatment may become necessary.

Practitioners and researchers must always rely on their own experience and knowledge in
evaluating and using any information, methods, compounds, or experiments described
herein. In using such information or methods they should be mindful of their own safety
and the safety of others, including parties for whom they have a professional responsibility.

To the fullest extent of the law, neither the Publisher nor the authors, contributors, or
editors, assume any liability for any injury and/or damage to persons or property as a
matter of products liability, negligence or otherwise, or from any use or operation of any
methods, products, instructions, or ideas contained in the material herein.

British Library Cataloguing in Publication Data
A catalogue record for this book is available from the British Library

Library of Congress Cataloging-in-Publication Data
A catalog record for this book is available from the Library of Congress

ISBN: 978-0-12-802318-1

For information on all Academic Press publications
visit our website at http://store.elsevier.com/

Working together
to grow libraries in
developing countries

www.elsevier.com • www.bookaid.org

CONTENTS

FOREWORD

Bent functions are fascinating mathematical objects. They were discovered by cryptographers who were searching for functions that are difficult to approximate by linear or affine functions. Bent functions are defined as functions that are at maximum distance to such weak functions. Bent functions were discovered independently by cryptographers in the US National Security Agency and in the Soviet Union; both nations decided to classify the results as confidential.

After one decade, Rothaus was allowed to publish his groundbreaking paper; it appeared in a journal on combinatorics in 1976. Around the same time (in 1972), Dillon published his seminal PhD thesis on elementary Hadamard difference sets. The first application area of bent functions considered in the open literature was coding theory. Academic cryptographers established the relation to cryptography only in 1989, when Meier and Staffelbach studied linear approximations of Boolean functions used in stream ciphers. This work stimulated broader interest in the topic, and inspired the author of this foreword to make some very modest contributions. Perhaps the largest impact on modern cryptography to date would be generated by the study of generalizations to vector Boolean functions that offer strong resistance against differential and linear attacks by Nyberg and others. This work resulted in the S-box used in the Advanced Encryption Standard (AES) that is today used in billions of devices. Other applications include wireless communications: sequences derived from bent functions can enhance code division multiple access (CDMA) transmission techniques.

Several years before Rothaus, Eliseev and Stepchenkov discovered bent functions in the USSR. Unfortunately their work is still classified as confidential. However, there is no doubt that the work of both authors inspired a large and valuable body of literature in Russian on the topic, some of which is public.

The author of this book has made a remarkable achievement. She has brought together the large body of knowledge on bent functions in both English and Russian in a single book. The book describes the history, presents definitions, and brings together all known results (125 theorems) and constructions in one integrated volume. It presents interesting perspectives based on the research of the author and a broad range of generalizations.

The literature in the first half century of bent functions is so vast that it is not possible to include the proofs. The book also contains many difficult open problems, enough to fill the careers of many mathematicians and cryptographers.

I hope that this book will inspire many researchers to explore the fascinating world of bent functions and to make progress on the rich and intricate problems in this world. I also hope that this book will increase mutual respect and understanding between researchers from the East and West and that it will lead to fruitful collaborations.

Bart Preneel
March 2015

PREFACE

Bent functions deserve
our bent to study them...

This book is devoted to such objects of discrete mathematics as Boolean *bent functions*. These functions have a remarkable property: each of them is at the maximal possible Hamming distance from the class of all affine Boolean functions. This extremal property distinguishes bent functions as the special mysterious class and leads to numerous applications of bent functions in combinatorics, coding theory, and cryptography.

Bent functions were introduced by O. Rothaus, an American mathematician, in the 1960s. At the same time, bent functions were studied in the USSR by V. A. Eliseev and O. P. Stepchenkov: they called such functions *minimal functions*. A little later J. A. Maiorana, R. L. McFarland, and J. Dillon proposed the first constructions of bent functions.

It was the early beginning...

The main goal of this book is to provide an overview of how the theory of bent functions developed from that time to this moment. This theory is still far from complete since too many questions remain open. We offer the most complete survey on bent functions. More than 125 theorems related to bent functions are included, and more than 400 references on bent functions are cited—from the very famous to very rare and widely unknown before. The book contains exclusive photographs of the first researchers in bent functions—most of them were never published before. Because of the large amount of work, not all important results are listed with the necessary details; some results are only mentioned, and we apologize for this limitation beforehand.

This book starts with basic definitions and historical aspects of the invention of bent functions. Applications of bent functions in cryptography (S-box construction, CAST, Grain, and HAVAL), discrete mathematics (Hadamard matrices, graphs, Kerdock codes, and bent codes) and communications (code division multiple access, bent sequences, and constant-amplitude codes) are discussed. We study basic properties of bent functions (degree restriction, affine transformations, rank, and duality) and equivalent representations of them (difference sets, designs, linear spreads, sets of subspaces, strongly regular graphs, and bent rectangles). Classifications of bent functions in a small number of variables are studied in detail (extended affine classification, classification in terms of bent rectangles, and graph classification for quadratic functions). An overview of algorithms for the generation of bent functions is presented.

Then we discuss combinatorial constructions of bent functions (simple iterative constructions, Maiorana–McFarland construction, partial spreads, Dillon's and Dobbertin's bent functions, minterm bent functions, and bent iterative functions). Then we come to relatively new algebraic constructions (Gold, Dillon, Kasami, Canteaut-Leander, and Canteaut-Charpin-Kuyreghyan bent exponents, and Niho bent functions) and discuss an algebraic approach in general. Connections between bent functions and other cryptographic properties (such as balancedness, correlation, and algebraic immunities) are also considered, together with some vectorial extensions.

Distances between bent functions are studied (minimal Hamming distance between bent functions, bounds on the number of bent functions at the minimal distance from a given one, locally metrical equivalence of bent functions, and the graph of minimal distances of bent functions). The group of automorphisms of the set of bent functions is established (it is proven that there are no other isometric mappings distinct from affine transformations that save the bent property of a function). Duality between the definitions of bent and affine functions is discussed.

Bounds on the number of bent functions are considered in detail. In our area of interest there are the best bounds for the number of bent functions up to 16 variables; for an arbitrary n, there is the best upper bound of C. Carlet and A. Klapper, and the best direct and iterative lower bounds of S. Agievich and the author, respectively. Hypotheses on the asymptotic value of the number of all bent functions are discussed. In connection with them the following question arises: Is it true that every Boolean function of degree up to $n/2$ can be represented as the sum of two bent functions? We consider this "bent sum decomposition problem" too, and prove that every Boolean function in n variables of a constant degree (less than or equal to $n/2$) can be represented as the sum of a constant number of bent functions in n variables.

Generalizations of bent functions with respect to their algebraic, combinatorial, and cryptographic properties are becoming more numerous and more widely studied from year to year. It is quite difficult not only to determine connections between generalizations, but also to collect information about all of them and provide a brief overview of the progress in this area. That is why a large part of this book is devoted to this theme. A systematic survey of the existing generalizations of bent functions and their known special subclasses is provided. Whenever possible we try to establish relations between various generalizations. We divide the generalizations into three groups: algebraic, combinatorial, and cryptographic. In the first group, we study q-valued bent functions, p-ary bent functions, bent functions over a finite field, generalized Boolean bent functions of Schmidt, bent functions from a finite Abelian group into the set of complex numbers on the unit circle, bent functions from a finite Abelian group into a finite Abelian group, non-Abelian bent functions, vectorial G-bent functions, and multidimensional bent functions on a finite Abelian group. In the second group, we deal with such generalizations and subclasses of bent functions as symmetric bent functions, homogeneous bent functions, rotation-symmetric bent functions, normal bent functions, self-dual and anti-self-dual bent functions, partially defined bent functions, plateaued functions, $\mathbb{Z}$-bent functions, and quantum bent functions. For the third, cryptographic, group in the sphere of our interest, there are semibent functions, balanced bent functions, partially bent functions, hyperbent functions, bent functions of higher order, and k-bent functions.

A large index completes the book. In general there are no proofs in the book: the huge volume of the results reviewed does not allow their inclusion. Moreover, we guess that there is no necessity in having proofs in such a book as this since many proofs are rather too special and will "slacken the pace" of an overview. There are only several proofs obtained by the author (automorphism group, bent iterative functions, etc.). But related to every result in this book we always include a reference to the original source. Thus, the interested reader can find all necessary details about the proofs.

I am very grateful to Mikhael M. Glukhov and Alexander V. Cheremushkin for their valuable advice related to this book and for providing me with several exclusive photographs of the first researchers of bent functions. I would like to honor the memory of Alexander A. Nechaev and express my deep gratitude to him for valuable discussions and support of the idea to write this book. My deep thanks go to Igor G. Shaposhnikov for providing me with the photographs of V. A. Eliseev and O. P. Stepchenkov. I express my gratitude to Sergey V. Agievich for several useful discussions. My kind thanks go to Alexander A. Evdokimov for his active support and a friendly atmosphere during the work. Finally, I thank Anastasia Gorodilova and Nikolay Kolomeec for their kind attention to this book and helpful remarks. This book was supported by the Sobolev Institute of Mathematics, Novosibirsk State University, and RFBR grants (projects 14-01-00507, 15-07-01328).

Finally, I wish good luck and inspiration to every researcher who is going to solve hard problems in bent functions or who is just thinking about this at the moment. Who knows, maybe bent functions are your bent!

Natalia Tokareva
Akademgorodok, Novosibirsk, Russia
February, 2015

NOTATION

p	a prime number (in most cases $p = 2$)
n	a natural number (usually even)
$\mathbb{F}_p$	the prime field, $\mathbb{F}_p = \{0, 1, \ldots, p - 1\}$
$\mathbb{F}_p^n$	the n-dimensional vector space over $\mathbb{F}_p$
$\mathbb{F}_{p^n}$	the finite field with p^n elements (also denoted $GF(p^n)$)
$\mathbb{F}_{p^n}^*$	the set of all nonzero elements of the field $\mathbb{F}_{p^n}$
$\mathrm{Aut}(\mathbb{F}_{p^n})$	the *Galois group* of the field $\mathbb{F}_{p^n}$; that is, the group of all its automorphisms with respect to superposition
$|M|$	the size of the set M
$\gcd(a, b)$	the *greatest common divisor* of two numbers a and b
$\oplus$	the sum over $\mathbb{F}_2$ (XOR operation)
$x = (x_1, \ldots, x_n)$	a binary vector over $\mathbb{F}_2$ of length n
$x \oplus y$	the sum of two binary vectors over $\mathbb{F}_2$, $x \oplus y = (x_1 \oplus y_1, \ldots, x_n \oplus y_n)$
$\langle x, y \rangle$	the standard *inner product* of vectors, where $\langle x, y \rangle = x_1 y_1 \oplus \cdots \oplus x_n y_n$
$x \preccurlyeq y$	the *precedence relation*: $x \preccurlyeq y$ if and only if for all $i = 1, \ldots, n$ $x_i \leqslant y_i$ holds (i.e., *x is covered* by y)
$d(x, y)$	the *Hamming distance* between vectors x and y
$\mathrm{wt}(x)$	the *Hamming weight* of a vector x
$\mathrm{wt}(k)$	the *Hamming weight* of a number k; that is, the Hamming weight of its binary representation
$f : \mathbb{F}_2^n \to \mathbb{F}_2$	a *Boolean function* in n variables
$F : \mathbb{F}_2^n \to \mathbb{F}_2^m$	a *vectorial Boolean function* in n variables
$\deg(f)$	the *degree* of a Boolean function
$\mathrm{ANF}(f)$	the *algebraic normal form* of a Boolean function
E^n	a *Boolean cube* of dimension n
$\mathrm{supp}(f)$	the *support* of a Boolean function f, where $\mathrm{supp}(f) = \{x \in \mathbb{F}_2^n : f(x) = 1\}$
$\mathrm{dist}(f, g)$	the *Hamming distance* between functions f and g; that is, $\mathrm{dist}(f, g) = |\{x \in \mathbb{F}_2^n : f(x) \neq g(x)\}|$
$\mathrm{wt}(f)$	the *Hamming weight* of a function f, $\mathrm{wt}(f) = |\mathrm{supp}(f)|$;
$\mathrm{tr}(c)$	a *trace* function, $\mathrm{tr} : \mathbb{F}_{2^n} \to \mathbb{F}_2$, defined as $\mathrm{tr}(c) = c + c^2 + c^{2^2} + c^{2^3} + c^{2^4} + \cdots + c^{2^{n-1}}$;
$\mathrm{tr}_k^n(c)$	a *trace* function, $\mathrm{tr}_k^n : \mathbb{F}_{2^n} \to \mathbb{F}_{2^k}$, defined as $\mathrm{tr}_k^n(c) = c + c^{2^k} + c^{2^{2k}} + c^{2^{3k}} + \cdots + c^{2^{k(n/k-1)}}$
N_f	*nonlinearity* of a Boolean function; that is, $N_f = \min_{a \in \mathbb{F}_2^n, b \in \mathbb{F}_2} \mathrm{dist}(f, \ell_{a,b})$, where $\ell_{a,b}$ is affine
$W_f(y)$	the *Walsh-Hadamard coefficient* of a Boolean function

$\widetilde{f}$	a *dual* bent function to a bent function f
$\mathcal{B}_n$	the set of all bent functions in n variables
$\mathcal{BI}_n$	the set of all bent iterative functions in n variables
$\mathrm{Aut}(\mathcal{M})$	the *group of automorphisms* of a subset M of Boolean functions
$\mathrm{GA}(n)$	the *general affine group*
$G_f = G(\mathbb{F}_2^n, \mathrm{supp}(f))$	a *Cayley graph* of a Boolean function; there is an edge between x and y if $x \oplus y$ belongs to $\mathrm{supp}(f)$
$\mathcal{PS}$	*partial spread* bent functions

CHAPTER 1

Boolean Functions

INTRODUCTION

In this chapter, we start with basic definitions related to Boolean functions. We consider the algebraic normal form of a Boolean function and the representation of a Boolean function over the Boolean cube. Extended affinely equivalent Boolean functions are defined as is the Walsh-Hadamard transform of a Boolean function. The finite field over $\mathbb{F}_2$ and its automorphisms are considered. It is shown how to associate Boolean functions in n variables with functions over the field $\mathbb{F}_{2^n}$. We discuss polynomial representations of Boolean and vectorial Boolean functions. Representations of a Boolean function in the trace form and in the reduced trace form are given. Some details on the degree of a Boolean function in the trace form and on monomial functions are presented. The notions introduced in this chapter will be useful throughout the book.

1.1 DEFINITIONS

Let $\mathbb{F}_2^n$ denote the n-dimensional vector space over the prime field $\mathbb{F}_2$. Let $x = (x_1, \ldots, x_n)$ be a vector over $\mathbb{F}_2$ of length n.

A *Boolean function in n variables* is an arbitrary function from $\mathbb{F}_2^n$ to $\mathbb{F}_2$. It is called Boolean in honor of the British mathematician and philosopher George Boole (1815-1864).

Every Boolean function can be defined by its *truth table*:

$x_1 \ldots x_n$	$f(x_1, \ldots, x_n)$
$0 \ldots 0$	$*$
$\vdots$	$\vdots$
$1 \ldots 1$	$*$

where in the first column there are all possible vectors of $\mathbb{F}_2^n$ and in the second column there are concrete values of a Boolean function taken on these vectors (denoted here by $*$). We suppose that the arguments of a function (i.e., vectors of length n) follow in lexicographical order. For

Bent Functions
http://dx.doi.org/10.1016/B978-0-12-802318-1.00001-7

example, if $n = 3$, the order is $(000), (001), (010), (011), (100), (101),$ $(110), (111)$.

For instance, the following are Boolean functions:

$g : \mathbb{F}_2^2 \to \mathbb{F}_2$ such that $g(00) = g(11) = 1$, $g(01) = g(10) = 0$;

$h : \mathbb{F}_2^3 \to \mathbb{F}_2$ such that $h(x) = 1$ if and only if x has two nonzero coordinates.

Their truth tables are as follows:

$x_1 x_2$	$g(x_1, x_2)$
0 0	1
0 1	0
1 0	0
1 1	1

$x_1 x_2 x_3$	$h(x_1, x_2, x_3)$
0 0 0	0
0 0 1	0
0 1 0	0
0 1 1	1
1 0 0	0
1 0 1	1
1 1 0	1
1 1 1	0

It is easy to count the number of Boolean functions. There are 2^{2^n} of them: to construct a function, one chooses 2^n values (0 or 1) for $f(x)$ when x runs through $\mathbb{F}_2^n$.

Every Boolean function in n variables can be uniquely determined by its *vector of values* of length 2^n. This is the transposed second column of its truth table.

In our examples, (1001) and (00010110) are vectors of values for g and h, respectively.

A *vectorial Boolean function* F in n variables is a function from $\mathbb{F}_2^n$ to $\mathbb{F}_2^m$, where m is an integer. It is also called an (n, m)-*function*. In what follows in this book, we consider $m = n$ unless otherwise stated. For vectorial Boolean functions we use uppercase letters, whereas Boolean functions are denoted with lowercase letters.

Every vectorial Boolean function in n variables can be presented as

$$F(x) = (f_1(x), \ldots, f_n(x)), \quad x \in \mathbb{F}_2^n,$$

where $f_1, \ldots, f_n$ are Boolean functions in n variables called *coordinate functions* of F. An arbitrary nonempty linear combination of coordinate functions is called a *component function* of a vectorial function F. In terms of the inner product, which will be introduced in the next section, a component function is a function $f^v(x) = \langle F(x), v \rangle$ for any nonzero $v \in \mathbb{F}_2^n$. In particular, every coordinate function is a component function.

For example, $F : \mathbb{F}_2^3 \to \mathbb{F}_2^3$ given by the truth table

$x_1 x_2 \, x_3$	$F(x_1, x_2, x_3)$
0 0 0	0 0 1
0 0 1	1 0 1
0 1 0	0 1 0
0 1 1	1 0 0
1 0 0	0 0 0
1 0 1	1 0 0
1 1 0	0 0 1
1 1 1	1 1 0

is a vectorial Boolean function with coordinate functions $f_1 = (01010101)$, $f_2 = (00100001)$, and $f_3 = (11000010)$; we list here vectors of values of them. Note that the component functions of F are $f_1, f_2, f_3, f_1 \oplus f_2 = (01110100), f_1 \oplus f_3 = (10010111), f_2 \oplus f_3 = (11100011),$ and $f_1 \oplus f_2 \oplus f_3 = (10110110)$.

1.2 ALGEBRAIC NORMAL FORM

Let $\oplus$ denote the addition modulo 2 (XOR). It is known that any Boolean function can be uniquely represented by its *algebraic normal form* (ANF):

$$f(x_1, \ldots, x_n) = \left(\bigoplus_{k=1}^{n} \bigoplus_{i_1, \ldots, i_k} a_{i_1, \ldots, i_k} \, x_{i_1} \cdot \ldots \cdot x_{i_k} \right) \oplus a_0,$$

where for each k indices $i_1, \ldots, i_k$ are pairwise distinct and sets $\{i_1, \ldots, i_k\}$ are exactly all different nonempty subsets of the set $\{1, \ldots, n\}$; coefficients $a_{i_1, \ldots, i_k}, a_0$ take values from $\mathbb{F}_2$. In the Russian mathematical literature, the ANF is usually called a *Zhegalkin polynomial* in honor of Ivan Zhegalkin (1869-1947), a Soviet mathematician who introduced such a representation in 1927.

For a Boolean function f, the number of variables in the longest item of its ANF is called the *algebraic degree* of a function (or briefly *degree*) and is denoted by deg(f). A Boolean function is *affine, quadratic, cubic,* and so on if its degree is not more than 1, or is equal to 2, 3, and so on, respectively.

For example, functions g and h given above have the ANFs $g(x_1, x_2) = x_1 \oplus x_2 \oplus 1$ and $h(x_1, x_2, x_3) = x_1 x_2 x_3 \oplus x_1 x_2 \oplus x_1 x_3 \oplus x_2 x_3$ and degrees 1 and 3, respectively.

On the set of all binary vectors the *precedence relation* can be defined: $x \preccurlyeq y$ if and only if for all $i = 1, \ldots, n$, $x_i \leqslant y_i$ (it is possible to say that x is *covered* by y). By using this relation, we can represent an ANF as follows:

$$f(x) = \bigoplus_{y \in \mathbb{F}_2^n} f_y x_1^{y_1} \cdot \cdots \cdot x_n^{y_n}, \quad \text{where } f_y = \bigoplus_{z \in \mathbb{F}_2^n, z \preccurlyeq y} f(z).$$

Here by x_i^δ we mean x_i if $\delta = 1$ and 1 if $\delta = 0$.

The *inner product* $\langle x, y \rangle$ of two binary vectors x and y of length n is equal to $x_1 y_1 \oplus \cdots \oplus x_n y_n$. Two vectors are *orthogonal* if $\langle x, y \rangle = 0$. Any affine Boolean function in n variables $x_1, \ldots, x_n$ can be represented as $\langle a, x \rangle \oplus b$ for an appropriate vector a and a binary constant b.

1.3 BOOLEAN CUBE AND HAMMING DISTANCE

In connection with Boolean functions, we consider the special graph *Boolean cube* denoted by E^n. Its vertices are all binary vectors of length n; there is an edge between vectors x and y if they differ in exactly one coordinate. The number n is called the *dimension* of a Boolean cube. Below one can see two-dimensional and three-dimensional Boolean cubes.

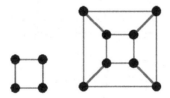

A four-dimensional cube is more complicated:

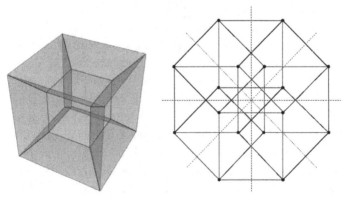

It can be better seen in following figure:

All vectors with exactly k nonzero coordinates form the k-layer of the Boolean cube. The size of the k-layer is $\binom{n}{k}$.

There are two useful interpretations of Boolean functions in the Boolean cube. First, a Boolean function in n variables can be associated with its vector of values, a vertex of the 2^n-dimensional Boolean cube. We denote such a vector for a Boolean function f by f: every time it will be clear from the context what we mean.

On the other hand, every Boolean function f in n variables can be associated with its *support*:

$$\mathrm{supp}(f) = \{x \in \mathbb{F}_2^n : f(x) = 1\}.$$

Obviously, supports of distinct Boolean functions are distinct too. Since any subset of $\mathbb{F}_2^n$ is the support of a certain Boolean function, there is a one-to-one correspondence between Boolean functions in n variables and subsets of $\mathbb{F}_2^n$.

The *Hamming distance* between two binary vectors x and y of length n is the number of coordinates in which they differ. We denote it by $d(x, y)$. The *Hamming weight* $\mathrm{wt}(x)$ of a binary vector x is the number of its nonzero coordinates. It is clear that $d(x, y) = \mathrm{wt}(x \oplus y)$.

We denote by $\mathrm{dist}(f, g)$ the *Hamming distance* between two Boolean functions f and g; it is the number of positions in which their vectors of values differ:

$$\mathrm{dist}(f, g) = |\{x \in \mathbb{F}_2^n : f(x) \neq g(x)\}|.$$

We denote by $\mathrm{wt}(f)$ the *Hamming weight* of a Boolean function f.

Let wt(k) denote the Hamming weight of an integer k—that is, the Hamming weight of its binary representation.

For instance, wt(19) = 3, since 19 can be represented by the binary vector (10011).

1.4 EXTENDED AFFINELY EQUIVALENT FUNCTIONS

One of the most general approaches to classification of Boolean functions is based on the notion of *affine equivalence* and *extended affine equivalence*.

Boolean functions f and g in n variables are *affinely equivalent* if there is a nondegenerate affine transformation of variables that maps one Boolean function to another. In other words, f and g are affinely equivalent if there is a nonsingular $n \times n$ matrix A and a vector b of length n such that

$$g(x) = f(Ax \oplus b) \quad \text{for every } x \in \mathbb{F}_2^n.$$

Boolean functions f and g in n variables are *extended affinely equivalent* if there is a nondegenerate affine transformation of variables that maps one Boolean function to another up to the addition of an affine function. So, f and g are extended affinely equivalent if there is a nonsingular $n \times n$ matrix A, vectors b and c of length n, and a constant $\lambda \in \mathbb{F}_2$ such that

$$g(x) = f(Ax \oplus b) \oplus \langle c, x \rangle \oplus \lambda \quad \text{for every } x \in \mathbb{F}_2^n.$$

Very often in the literature equivalence of this type is also called "affine equivalence"; so, when reading an article, we should be attentive and understand clearly every time what the authors mean by "affine equivalence."

In what follows, by *"equivalent Boolean functions"* we mean functions that are extended affinely equivalent unless stated otherwise.

For instance, Boolean functions $f(x_1, x_2, x_3, x_4) = x_1 x_2 \oplus x_3 x_4$ and $g(x_1, x_2, x_3, x_4) = x_1 x_2 \oplus x_1 x_3 \oplus x_2 x_4 \oplus x_3$ are extended affinely equivalent. In fact, $g(x) = f(Ax \oplus b) \oplus \langle c, x \rangle \oplus \lambda$, where

$$A = \begin{pmatrix} 1 & 0 & 0 & 1 \\ 0 & 1 & 1 & 0 \\ 0 & 0 & 1 & 0 \\ 0 & 0 & 0 & 1 \end{pmatrix}, \quad b = (0011), \quad c = (0001), \quad \lambda = 1.$$

In detail. $f(Ax \oplus b) \oplus \langle c, x \rangle \oplus \lambda = f(x_1 \oplus x_4, x_2 \oplus x_3, x_3 \oplus 1, x_4 \oplus 1) \oplus x_4 \oplus 1 = (x_1 \oplus x_4)(x_2 \oplus x_3) \oplus (x_3 \oplus 1)(x_4 \oplus 1) \oplus x_4 \oplus 1 = x_1 x_2 \oplus x_1 x_3 \oplus x_2 x_4 \oplus x_3 = g(x).$

The street in Cork, Ireland, where George Boole lived from 1849 to 1855. His house is the last on the right

1.5 WALSH-HADAMARD TRANSFORM

In this section, we give a definition that is very important with respect to bent functions. The *Walsh-Hadamard transform* of a Boolean function f in n variables is the integer-valued function on $\mathbb{F}_2^n$ defined as

$$W_f(y) = \sum_{x \in \mathbb{F}_2^n} (-1)^{\langle x,y \rangle \oplus f(x)} \quad \text{for every } y \in \mathbb{F}_2^n.$$

Numbers $W_f(y)$ are called *Walsh-Hadamard coefficients* of a Boolean function f. The ordered multiset

$$W_f = \{W_f(x) : x \in \mathbb{F}_2^n\},$$

where vector x runs through $\mathbb{F}_2^n$ in lexicographical order, is the *Walsh-Hadamard spectrum* of a function f.

For instance, if $g(x_1, x_2) = x_1 x_2$, then

$$W_g(00) = (-1)^{\langle 00,00 \rangle \oplus 0} + (-1)^{\langle 01,00 \rangle \oplus 0} + (-1)^{\langle 10,00 \rangle \oplus 0}$$

$$+(-1)^{\langle 11,00 \rangle \oplus 1} = 2,$$

$$W_g(01) = (-1)^{\langle 00,01 \rangle \oplus 0} + (-1)^{\langle 01,01 \rangle \oplus 0} + (-1)^{\langle 10,01 \rangle \oplus 0}$$

$$+(-1)^{\langle 11,01 \rangle \oplus 1} = 2,$$

$$W_g(10) = (-1)^{\langle 00,10 \rangle \oplus 0} + (-1)^{\langle 01,10 \rangle \oplus 0} + (-1)^{\langle 10,10 \rangle \oplus 0}$$

$$+(-1)^{\langle 11,10 \rangle \oplus 1} = 2,$$

$$W_g(11) = (-1)^{\langle 00,11 \rangle \oplus 0} + (-1)^{\langle 01,11 \rangle \oplus 0} + (-1)^{\langle 10,11 \rangle \oplus 0}$$

$$+(-1)^{\langle 11,11 \rangle \oplus 1} = -2.$$

Thus, the Walsh-Hadamard spectrum of g is $\{2, 2, 2, -2\}$.

It is not difficult to prove that the Walsh-Hadamard spectrum determines a Boolean function in a unique manner.

Theorem 1. *For a Boolean function f in n variables and an arbitrary vector x of length n,*

$$(-1)^{f(x)} = \frac{1}{2^n} \sum_{y \in \mathbb{F}_2^n} W_f(y)(-1)^{\langle x,y \rangle}.$$

Walsh-Hadamard coefficients satisfy *Parseval's equality*.

Theorem 2. *For a Boolean function f in n variables, Parseval's equality holds:*

$$\sum_{y \in \mathbb{F}_2^n} \left(W_f(y) \right)^2 = 2^{2n}.$$

As far as the number of all coefficients $W_f(y)$ of a Boolean function f is 2^n, we have immediately the following theorem:

Theorem 3. *For every Boolean function f in n variables,*

$$\max_{y \in \mathbb{F}_2^n} |W_f(y)| \geqslant 2^{n/2}.$$

1.6 FINITE FIELD AND BOOLEAN FUNCTIONS

Let $\mathbb{F}_{2^n}$ be the finite field of order 2^n. It is denoted also as GF(2^n) in honor of the French mathematician Évariste Galois (1811-1832). As usual, the set of all nonzero elements of the field is denoted by $\mathbb{F}_{2^n}^*$. Recall that $\mathbb{F}_{2^n}^*$ is a cyclic group of order $2^n - 1$ with respect to multiplication; its generating element is called a *primitive element* of the field $\mathbb{F}_{2^n}$.

The vector space $\mathbb{F}_2^n$ can be endowed with the structure of the finite field. Consider a way how to do it.

Let x be a formal symbol here, a *variable*, not belonging to $\mathbb{F}_2$. Recall that a *polynomial of degree d* over $\mathbb{F}_2$ is a formal expression

$$a(x) = a_1 x^d + a_2 x^{d-1} + \cdots + a_{d-1} x^2 + a_d x + a_{d+1},$$

where all coefficients a_i are from $\mathbb{F}_2$ (and $a_1 \neq 0$). Polynomials over $\mathbb{F}_2$ can be summed and multiplied in the natural way. A polynomial is *irreducible* if it cannot be represented as the product of two polynomials of smaller degrees.

For instance, $x^3 + x + 1$ is irreducible, whereas $x^3 + 1$ is not irreducible since it is equal to $(x + 1)(x^2 + x + 1)$.

Let $g(x)$ be an irreducible polynomial of degree n. For a binary vector $c = (c_1, \ldots, c_n)$ of length n, we put into the correspondence the polynomial $c(x)$ of degree $n - 1$ over $\mathbb{F}_2$ of the form

$$c(x) = c_1 x^{n-1} + c_2 x^{n-2} + \cdots + c_{n-1}x + c_n.$$

We define summing and multiplying of vectors from $\mathbb{F}_2^n$ as follows:

$$c + c' = d, \text{ where } d(x) = c(x) + c'(x);$$
$$c \cdot c' = d, \quad \text{where } d(x) = c(x) \cdot c'(x) \mod g(x).$$

Then $\mathbb{F}_2^n$ endowed with these operations is a finite field—namely, $\mathbb{F}_{2^n}$. The polynomial $g(x)$ is called a *generator polynomial* of the field.

Further, we identify a Boolean function in n variables with a function from $\mathbb{F}_{2^n}$ to $\mathbb{F}_2$; a vectorial Boolean function in n variables with a function from $\mathbb{F}_{2^n}$ to $\mathbb{F}_{2^n}$. We also call these functions *Boolean* and *vectorial Boolean*, respectively. If it is not important, we do not concretize an isomorphism between vector space $\mathbb{F}_2^n$ and the field $\mathbb{F}_{2^n}$.

An *automorphism* of the finite field $\mathbb{F}_{2^n}$ over $\mathbb{F}_2$ is a one-to-one mapping φ of $\mathbb{F}_{2^n}$ to itself such that
(1) φ leaves elements of $\mathbb{F}_2$ without changes;
(2) φ preserves operations in the field—that is, for any $a, b \in \mathbb{F}_{2^n}$,

$$\varphi(a + b) = \varphi(a) + \varphi(b),$$

$$\varphi(a \cdot b) - \varphi(a) \cdot \varphi(b).$$

The *Galois group* of the field $\mathbb{F}_{2^n}$ is the group of all its automorphisms with respect to operation superposition. Usually it is called the *automorphism group* of $\mathbb{F}_{2^n}$ and is denoted as $\mathrm{Aut}(\mathbb{F}_{2^n})$.

Theorem 4. *The automorphism group of the field $\mathbb{F}_{2^n}$ is a cyclic group of order n with a generating element $\varphi : a \to a^2$.*

Thus, $\mathrm{Aut}(\mathbb{F}_{2^n}) = \{e, \varphi, \varphi^2, \ldots, \varphi^{n-1}\}$, where e is the identity map.

1.7 TRACE FUNCTION

Recall that a *trace* is a function over a finite field $\mathbb{F}_{2^n}$, defined as follows:

$$\mathrm{tr}(c) = c + c^2 + c^{2^2} + c^{2^3} + c^{2^4} + \cdots + c^{2^{n-1}}.$$

The most important property of the trace is that its values belong to the prime subfield $\mathbb{F}_2$. Some other properties are briefly listed here:

- For any c, $\mathrm{tr}^2(c) = \mathrm{tr}(c^2) = \mathrm{tr}(c)$.
- The trace is a linear function—that is, $\mathrm{tr}(c' + c'') = \mathrm{tr}(c') + \mathrm{tr}(c'')$ for all c' and c''.
- The trace is a balanced function—that is, it takes values 0 and 1 equally often.

Since there is an isomorphism between $\mathbb{F}_2^n$ and $\mathbb{F}_{2^n}$, it is possible to identify the trace function with a Boolean function in n variables.

Consider an example where $n = 3$. Let $\mathbb{F}_{2^3}$ be constructed with the irreducible polynomial $g(x) = x^3 + x + 1$. Element $x + 1$ is primitive (check that all its powers from the first to the sixth are distinct). We denote it by α. Below we show how to calculate the trace of each element of the field $\mathbb{F}_{2^3}$. In the last column, the vector of values of the Boolean function corresponding to $\mathrm{tr}(\cdot)$ is presented. Note that it is a linear function, $f(x_1, x_2, x_3) = x_3$.

Vector	Polynomial	α^k	Trace $\mathrm{tr}(c) = c + c^2 + c^4$	f
(000)	0	—	0	0
(001)	1	α^0	$\alpha^0 + \alpha^{0 \cdot 2} + \alpha^{0 \cdot 4} = 1$	1
(010)	x	α^5	$\alpha^5 + \alpha^{5 \cdot 2} + \alpha^{5 \cdot 4} = \alpha^5 + \alpha^3 + \alpha^6 = 0$	0
(011)	$x + 1$	α^1	$\alpha^1 + \alpha^{1 \cdot 2} + \alpha^{1 \cdot 4} = 1$	1
(100)	x^2	α^3	$\alpha^3 + \alpha^{3 \cdot 2} + \alpha^{3 \cdot 4} = \alpha^3 + \alpha^6 + \alpha^5 = 0$	0
(101)	$x^2 + 1$	α^2	$\alpha^2 + \alpha^{2 \cdot 2} + \alpha^{2 \cdot 4} = \alpha^2 + \alpha^4 + \alpha^1 = 1$	1
(110)	$x^2 + x$	α^6	$\alpha^6 + \alpha^{6 \cdot 2} + \alpha^{6 \cdot 4} = \alpha^6 + \alpha^5 + \alpha^3 = 0$	0
(111)	$x^2 + x + 1$	α^4	$\alpha^4 + \alpha^{4 \cdot 2} + \alpha^{4 \cdot 4} = \alpha^4 + \alpha^1 + \alpha^2 = 1$	1

In general, the trace function is linear for any isomorphism between $\mathbb{F}_2^n$ and $\mathbb{F}_{2^n}$. Moreover, the following theorem holds:

Theorem 5. *The set of all linear Boolean functions in n variables coincides with the set of functions $\ell_a(c) = \mathrm{tr}(a \cdot c)$, where a runs through the field $\mathbb{F}_{2^n}$.*

It is clear now that an isomorphism between $\mathbb{F}_2^n$ and $\mathbb{F}_{2^n}$ can be chosen in such a way that

$$\mathrm{tr}(c) = \langle a, c \rangle$$

for any preliminary fixed nonzero a from $\mathbb{F}_{2^n}$.

In fact, the trace function can be defined from $\mathbb{F}_{2^n}$ to an arbitrary subfield $\mathbb{F}_{2^k}$ (not only if $k = 1$). Note that k should divide n in this case. By definition,

$$\mathrm{tr}_k^n(c) = c + c^{2^k} + c^{2^{2k}} + c^{2^{3k}} + \cdots + c^{2^{k(n/k-1)}}.$$

In this notation, tr_1^n is the usual trace function tr.

1.8 POLYNOMIAL REPRESENTATION OF A BOOLEAN FUNCTION

Consider a very useful and powerful way to represent Boolean and vectorial functions as functions over the field $\mathbb{F}_{2^n}$.

Every vectorial Boolean function F in n variables that is a function from $\mathbb{F}_{2^n}$ to $\mathbb{F}_{2^n}$ can be uniquely represented in the univariate *polynomial form* (or *polynomial representation*) over $\mathbb{F}_{2^n}$ of degree not more than $2^n - 1$:

$$F(c) = \sum_{j=0}^{2^n-1} a_j c^j, \quad \text{where } a_j \in \mathbb{F}_{2^n}.$$

Indeed, the number of all vectorial Boolean functions in n variables is $(2^n)^{2^n}$. The number of distinct polynomials $\sum_{j=0}^{2^n-1} a_j x^j$ is again $(2^n)^{2^n}$. Note that distinct polynomials define distinct vectorial functions.

In what follows it is naturally assumed that $c^0 = 1$ for every $c \in \mathbb{F}_{2^n}$. Note that $c^{2^n-1} = 1$ for all $c \in \mathbb{F}_{2^n}^*$ and $c^{2^n-1} = 0$, if $c = 0$.

As far as $\mathbb{F}_2$ is a subfield of $\mathbb{F}_{2^n}$, it is clear that a Boolean function $f : \mathbb{F}_{2^n} \to \mathbb{F}_2$ is a partial case of a vectorial function. Hence, f can also be uniquely represented as a polynomial over $\mathbb{F}_{2^n}$—that is,

$$f(c) = \sum_{j=0}^{2^n-1} a_j c^j, \quad \text{where } a_j \in \mathbb{F}_{2^n}$$

for some appropriate a_j. Note that an arbitrary polynomial on the right side of this equation takes values from the subfield $\mathbb{F}_2$ if and only if a_0 and a_{2^n-1} belong to $\mathbb{F}_2$ and $a_{2i} = a_i^2$ for all $i = 1, \ldots, 2^n - 2$, where the index $2i$ should be taken modulo $2^n - 1$. Such a representation for a Boolean function is called the *polynomial form* (or *polynomial representation*) of a Boolean function. For more details, see [46].

Note that bivariate polynomial forms are also used.

1.9 TRACE REPRESENTATION OF A BOOLEAN FUNCTION

Let us discuss another representation of a Boolean function over the finite field called the *trace form* (or *trace representation*).

Consider a trace projection of a vectorial Boolean function F in n variables into a Boolean function f. Namely, let

$$f(c) = \mathrm{tr}\left(\sum_{j=0}^{2^n-1} a_j c^j\right), \qquad \text{where } c \in \mathbb{F}_{2^n}.$$

It is easy to see that an arbitrary Boolean function f can be obtained in this way.

Namely, prove that for an arbitrary Boolean function f there is always a vectorial Boolean function F such that $f(c) = \mathrm{tr}(F(c))$. Really, as far as the trace is a nonconstant linear function, there is a nonzero vector b such that for every c $\mathrm{tr}(c) = \langle b, c \rangle = b_1 c_1 \oplus \cdots \oplus b_n c_n$. Take any i such that $b_i = 1$. Consider a vectorial Boolean function $F : \mathbb{Z}_2^n \to \mathbb{Z}_2^n$ with coordinate functions $f_j(c) = 0$ if $j \neq i$, and $f_i(c) = f(c)$. Then obviously, $f(c) = \mathrm{tr}(F(c))$ for every c.

Using knowledge of the automorphism group of $\mathbb{F}_{2^n}$, we can easily prove that for any $a \in \mathbb{F}_{2^n}$ there is an element $d \in \mathbb{F}_{2^n}$ such that $\mathrm{tr}(ac^2) = \mathrm{tr}(dc)$ for all $c \in \mathbb{F}_{2^n}$. According to this fact, we can simplify the representation for f. Let us summarize this in the theorem below.

Recall that a *cyclotomic class* modulo $2^n - 1$ with a representative k is the set $\{k, 2k, 2^2 k, 2^3 k, \ldots\}$, where all the numbers are considered modulo $2^n - 1$. We denote by CS the set of minimal representatives of all cyclotomic classes modulo $2^n - 1$.

For instance, if $n = 4$, there are five cyclotomic classes—$\{0\}$, $\{1, 2, 4, 8\}$, $\{3, 6, 12, 9\}$, $\{5, 10\}$, and $\{7, 14, 13, 11\}$—with representatives 0, 1, 3, 5, and 7, respectively.

Thus, one comes to the theorem:

Theorem 6. *A Boolean function $f : \mathbb{F}_{2^n} \to \mathbb{F}_2$ can be represented in the trace form*

$$f(c) = \mathrm{tr}\left(\sum_{j \in CS} a_j c^j\right) + \mathrm{tr}(a_{2^n-1} c^{2^n-1}),$$

for the appropriate elements $a_j \in \mathbb{F}_{2^n}$.

This representation is called the *trace form* (or *trace representation*) of a Boolean function. As we will see in Theorem 9 that all coefficients a_j such that $\mathrm{wt}(j) > \deg(f)$ can be taken to be zero.

Note that the trace representation is not unique. But by placing restrictions on coefficients a_j, we can make it so. This question was discussed in [220]; we consider it after an example.

Consider an example of how to get a trace form of a Boolean function from its ANF and vice versa.

Let $n = 3$ and let field $\mathbb{F}_{2^3}$ be constructed using the irreducible polynomial $g(x) = x^3 + x + 1$ as a generator. Denote by α the primitive element $x + 1$ of the field. Here we work with the same $g(x)$ and primitive element as in Section 1.7.

Let us find a trace representation for a Boolean function $f_1(c_1, c_2, c_3) = c_1 \oplus c_2 \oplus c_3$, where $c_i \in \mathbb{F}_2$. Note that the function f_1 is linear, and hence by Theorem 5 its trace representation can be found as $\text{tr}(a \cdot c)$ for the appropriate $a \in \mathbb{F}_{2^3}$. Let us find this a. We note that $f_1(001) = 1$; this means that $\text{tr}(a \cdot 1)$ should be equal to 1. Since $\text{tr}(\alpha^3) = 0$ we realize that a cannot be equal to α^3, α^6, α^5, and 0 according to the property $\text{tr}(c^2) = \text{tr}(c)$. Then we use the next value of f_1—namely, $f_1(010) = 1$. Hence, $\text{tr}(a \cdot \alpha^5)$ should be equal to 1 since vector (010) corresponds to α^5; see the table in Section 1.7. Then we see that a cannot be equal to 1 and α since $\text{tr}(\alpha^5) = tr(\alpha^6) = 0$. Thus, only two possibilities remain: check $a = \alpha^2$ and $a = \alpha^4$. We see that $f_1(100) = 1$, and hence $\text{tr}(a \cdot \alpha^3)$ should be equal to 1. As far as $\text{tr}(\alpha^2 \cdot \alpha^3) = 0$, we conclude that a is not equal to α^2. Hence, $a = \alpha^4$ and $f_1(c) = \text{tr}(\alpha^4 \cdot c)$.

It is more difficult to find a trace representation for a nonlinear function. For example, let $f_2(c_1, c_2, c_3) = c_1 c_2$. Since $\text{CS} = \{0, 1, 3\}$ and $c^0 = 1$ for all $c \in \mathbb{F}_{2^3}$, then by Theorem 6 we can find a trace representation of f_2 that looks like $\text{tr}(a_0) + \text{tr}(a_1 c) + \text{tr}(a_3 c^3) + \text{tr}(a_7 c^7)$ for appropriate a_0, a_1, a_3, and a_7 from $\mathbb{F}_{2^3}$. Since $f_2(000) = 0$, a_0 can be taken to be zero. From the remark after Theorem 6 (see in detail Theorem 9) a_7 can be taken to be zero because $\text{wt}(7) = 3$, whereas $\deg(f_2) = 2$. After some testing of a_1 and a_3, one can find a variant for a trace representation—for example, $f_2(c) = \text{tr}(\alpha^4 c) + \text{tr}(c^3)$.

It is easier to find the ANF of a function if you know its trace representation. For example, let $f_3(c) = \text{tr}(\alpha) + \text{tr}(\alpha \cdot c^3) + \text{tr}(\alpha^2 \cdot c^7)$. We are looking for the ANF of the general view: $f_3(c_1, c_2, c_3) = d'_{1,2,3} c_1 c_2 c_3 \oplus d'_{1,2} c_1 c_2 \oplus \oplus d'_{1,3} c_1 c_3 \oplus d'_{2,3} c_2 c_3 \oplus d'_1 c_1 \oplus d'_2 c_2 \oplus d'_3 c_3 \oplus d'_0$. We can write vectors of values for all monomials in the trace representation—namely, $\text{tr}(\alpha) = (11111111)$, $\text{tr}(\alpha \cdot c^3) = (01110100)$, and $\text{tr}(\alpha^2 \cdot c^7) = (01111111)$. Then, $f_3 = (11110100)$. Further, step by step it is easy to find all the coefficients d'. Thus, $f_3(c) = c_1 c_2 c_3 \oplus c_1 c_3 \oplus c_1 \oplus 1$.

Consider the restrictions on a trace representation in order to make it unique for a Boolean function. Let j be an element of CS. Define

$$r(j) = \min\left\{t : \frac{(2^n - 1)}{\gcd(2^n - 1, j)} \text{ divides } 2^t - 1\right\}.$$

Let $\xi_j \in \mathbb{F}_{2^n}$ be an element such that $\text{tr}^n_{r(j)}(\xi) = e$.

Kuzmin et al. [220] have proven the following theorem:

Theorem 7. *A Boolean function $f : \mathbb{F}_{2^n} \to \mathbb{F}_2$ can be uniquely represented in the reduced trace form*

$$f(c) = \mathrm{tr}\left(\sum_{j \in CS} b_j \xi_j c^j\right) + \mathrm{tr}(b_{2^n-1} \xi_{2^n-1} c^{2^n-1}),$$

for the appropriate elements $b_j \in \mathbb{F}_{2^{r(j)}}$.

Such a unique representation of a Boolean function is called the *reduced trace form (representation)*. The important statement on the degree of a Boolean function in the reduced trace representation was also obtained in [220]:

Theorem 8. *For the degree of a Boolean function $f : \mathbb{F}_{2^n} \to \mathbb{F}_2$ represented in the reduced trace form (as in Theorem 7),*

$$\deg(f) = \max\{\mathrm{wt}(j) : j \in CS, b_j \neq 0\}.$$

Since the reduced trace representation is a particular case of a trace representation (i.e., the representation from Theorem 7 is a particular case of the representation from Theorem 6), then the following theorem holds:

Theorem 9. *For the degree of a Boolean function $f : \mathbb{F}_{2^n} \to \mathbb{F}_2$ represented in a trace form (as in Theorem 6),*

$$\deg(f) \leqslant \max\{\mathrm{wt}(j) : j \in CS, a_j \neq 0\}.$$

Moreover, a trace representation of a Boolean function of degree d can be constructed with coefficients a_j being zero if $\mathrm{wt}(j) > d$ since Theorems 7 and 8 guarantee the existence of such a representation.

The trace form is one of the most useful instruments for working with cryptographic Boolean functions. In Chapter 9, we discuss bent functions constructed in trace forms with additional special properties.

1.10 MONOMIAL BOOLEAN FUNCTIONS

Consider Boolean functions with the simplest trace representations: every such form contains only one monomial. A Boolean function f in n variables is called *monomial* if it can be represented as

$$f(c) = \mathrm{tr}(ac^d), \quad c \in \mathbb{F}_{2^n},$$

for the appropriate $a \in \mathbb{F}_{2^n}$ and integer d, where $1 \leqslant d \leqslant 2^n - 1$. If $d = 1$, the monomial functions coincide with all linear Boolean functions.

From Theorem 8 the next theorem follows:

Theorem 10. *The degree of an arbitrary nonzero monomial Boolean function* $f(c) = \text{tr}(ac^d)$, *where* $a \neq 0$, *is equal to the Hamming weight of the number* d.

A Boolean function $\text{tr}(ac^d)$ is called a *proper monomial* if d and $2^n - 1$ are coprime.

Monomial bent functions will be considered in Chapters 9 and 17.

Describe, for instance, all proper monomial Boolean functions if $n = 3$. *Let* $g(x) = x^3 + x + 1$ *be a generator polynomial of the field* $\mathbb{F}_{2^n}$. *As before, let* $\alpha = x + 1$ *be a primitive element.*

Then all linear functions can be represented in trace form like this:

Vector	a	$\text{tr}(c)$	$\text{tr}(ac)$	Value	$f(x_1, x_2, x_3)$
(000)	0	0	$\text{tr}(0)$	00000000	0
(001)	1	1	$\text{tr}(c)$	01010101	x_3
(010)	α^5	0	$\text{tr}(\alpha^5 c)$	00001111	x_1
(011)	α^1	1	$\text{tr}(\alpha^1 c)$	01011010	$x_1 \oplus x_3$
(100)	α^3	0	$\text{tr}(\alpha^3 c)$	00110011	x_2
(101)	α^2	1	$\text{tr}(\alpha^2 c)$	01100110	$x_2 \oplus x_3$
(110)	α^6	0	$\text{tr}(\alpha^6 c)$	00111100	$x_1 \oplus x_2$
(111)	α^4	1	$\text{tr}(\alpha^4 c)$	01101001	$x_1 \oplus x_2 \oplus x_3$

Consider also the case $d = 3$. *Since 3 has the Hamming weight 2, we are dealing with quadratic Boolean functions:*

Vector	a	$\text{tr}(ac^3)$	Value	$f(x_1, x_2, x_3)$
(000)	0	$\text{tr}(0)$	00000000	0
(001)	1	$\text{tr}(c^3)$	01101010	$x_1 x_2 \oplus x_1 \oplus x_2 \oplus x_3$
(010)	α^5	$\text{tr}(\alpha^5 c^3)$	00011110	$x_2 x_3 \oplus x_1$
(011)	α^1	$\text{tr}(\alpha^1 c^3)$	01110100	$x_1 x_2 \oplus x_2 x_3 \oplus x_2 \oplus x_3$
(100)	α^3	$\text{tr}(\alpha^3 c^3)$	00100111	$x_1 x_3 \oplus x_2 x_3 \oplus x_2$
(101)	α^2	$\text{tr}(\alpha^2 c^3)$	01001101	$x_1 x_2 \oplus x_1 x_3 \oplus x_2 x_3 \oplus x_1 \oplus x_3$
(110)	α^6	$\text{tr}(\alpha^6 c^3)$	00111001	$x_1 x_3 \oplus x_1 \oplus x_2$
(111)	α^4	$\text{tr}(\alpha^4 c^3)$	01010011	$x_1 x_2 \oplus x_1 x_3 \oplus x_3$

CHAPTER 2

Bent Functions: An Introduction

INTRODUCTION

In this chapter, three definitions of a *bent function* are given: via nonlinearity, by using Walsh–Hadamard coefficients, and in terms of derivatives. Nonlinearity of a random Boolean function is discussed. Several open problems in bent functions are included. In the last section, we list surveys on bent functions in separate articles and in chapters in books on discrete mathematics and cryptography.

2.1 DEFINITION OF A NONLINEARITY

The *nonlinearity* of a Boolean function f in n variables is the Hamming distance N_f from this function to the set of all affine functions—that is,

$$N_f = \min_{a \in \mathbb{F}_2^n, b \in \mathbb{F}_2} \text{dist}(f, \ell_{a,b}),$$

where $\ell_{a,b}(x) = \langle a, x \rangle \oplus b$ is an affine function. S. W. Golomb, in 1959, was one of the first researchers who introduced this parameter.

For instance, find the nonlinearity of $f(x_1, x_2, x_3) = x_1 \oplus x_2 x_3$. The vector of values of this function is $f = (00011110)$, of weight 4. Note that the weight of an arbitrary affine function in three variables is equal to 0, 4, or 8. Enumerate the vectors of values of all affine Boolean functions in three variables:

0	00000000	1	11111111
x_1	00001111	$x_1 \oplus 1$	11110000
x_2	00110011	$x_2 \oplus 1$	11001100
x_3	01010101	$x_3 \oplus 1$	10101010
$x_1 \oplus x_2$	00111100	$x_1 \oplus x_2 \oplus 1$	11000011
$x_1 \oplus x_3$	01011010	$x_1 \oplus x_3 \oplus 1$	10100101
$x_2 \oplus x_3$	01100110	$x_2 \oplus x_3 \oplus 1$	10011001
$x_1 \oplus x_2 \oplus x_3$	01101001	$x_1 \oplus x_2 \oplus x_3 \oplus 1$	10010110

Since f is not affine and is of even weight, $N_f \geqslant 2$. The distance 2 is reached, for example, between f and function x_1, so $N_f = 2$.

Bent Functions
http://dx.doi.org/10.1016/B978-0-12-802318-1.00002-9

2.2 NONLINEARITY OF A RANDOM BOOLEAN FUNCTION

In 1998, Olejár and Stanek [293] investigated cryptographic properties of a random Boolean function in n variables.

They have proven the following theorem:

Theorem 11. *There is a constant c such that if n is big enough, then for almost every Boolean function f in n variables*

$$N_f \geqslant 2^{n-1} - c\sqrt{n}\,2^{n/2}.$$

In 2002, this fact was independently obtained by Carlet [40]. Here by "almost every" we understand "with probability tending to 1."

Let the nonlinearity of an arbitrary Boolean function g in n variables be expressed in the form $N_g = 2^{n-1} - S(g)$, where $S(g)$ is a certain function. In 2006, Rodier [315] determined the asymptotic value of the nonlinearity of a Boolean function.

Let V^∞ be the space of infinite sequences of elements of $\mathbb{F}_2$ which are almost all equal to zero. Let $f : V^\infty \to \mathbb{Z}_2$. Denote by f_n a restriction of f on the set $\mathbb{F}_2^n$. Rodier [315] has proven the following theorem:

Theorem 12. *For almost every Boolean function $f : V^\infty \to \mathbb{F}_2$,*

$$\lim_{n \to \infty} \frac{S(f_n)}{2^{n/2}\sqrt{n}} = \sqrt{2\ln 2}.$$

Thus, we see that the bigger n is, the higher is the nonlinearity of a random function in n variables; moreover, as we will see later, this nonlinearity becomes closer to the nonlinearity of a bent function!

It is known that for cryptographic applications a Boolean function should have many other properties. We can say that Theorem 12 "guarantees" the possibility to choose a function of high nonlinearity without significant restrictions on other properties. But, of course, this does not mean that such a function can be easily constructed. Similar "paradoxes" already exist in Boolean functions—for example, in complexity theory.[1] In our case, the asymptotic bound of Theorem 12 gives a certain *level* of nonlinearity of a "not bad" cryptographic Boolean function [316].

2.3 DEFINITION OF A BENT FUNCTION

Nonlinearity of any Boolean function in n variables can be found by using Walsh-Hadamard coefficients. Since

---

[1]C. Shannon has proven that almost every Boolean function has a very large complexity of realization; asymptotically, it equals the complexity of the "most complicated" function. But until now no example of such a function has been constructed.

$$\mathrm{dist}(f, \ell_{a,0}) = 2^{n-1} - \frac{1}{2} W_f(a),$$

$$\mathrm{dist}(f, \ell_{a,1}) = 2^{n-1} + \frac{1}{2} W_f(a),$$

one can easily get the following theorem:

Theorem 13. *Let f be a Boolean function in n variables. Then,*

$$N_f = 2^{n-1} - \frac{1}{2} \max_{y \in \mathbb{F}_2^n} |W_f(y)|.$$

From Theorems 3 and 13, it follows that $N_f \leqslant 2^{n-1} - 2^{(n/2)-1}$.

A Boolean function f in n variables is called *maximal nonlinear* if parameter N_f takes its maximal possible value.

Theorem 14. *If n is even, then the maximal possible value of nonlinearity is* $2^{n-1} - 2^{(n/2)-1}$.

If n is odd, this maximal possible value is still unknown.

Definition 1. A *bent function* is a maximal nonlinear Boolean function with an even number of variables. Its nonlinearity is $2^{n-1} - 2^{(n/2)-1}$.

By Theorems 2 and 3, the maximal nonlinearity of a Boolean function can be achieved only with equal absolute values of all coefficients $W_f(y)$—namely, $|W_f(y)| = 2^{n/2}$. Thus, we come to the following equivalent definition.

Definition 2. A *bent function* is a Boolean function in n variables (n is even) such that $W_f(y) = \pm 2^{n/2}$ for every y.

Following to approach of Meier and Staffelbach [262], we can define a bent function as a *perfect nonlinear function with respect to linear structures*. Namely, we have the following definition:

Definition 3. A *bent function* is a Boolean function in n variables (n is even) such that for any nonzero vector y its derivative $D_y f(x) = f(x) \oplus f(x \oplus y)$ is balanced—that is, it takes values 0 and 1 equally often.

Denote by $\mathcal{B}_n$ the set of all bent functions in n variables.

Consider several examples. On the figure bellow you can see a Boolean cube of dimension 4 in which black and white vertices are bent and affine functions, respectively. Since $n = 2$, every bent function in two variables is on the distance $2^{n-1} - 2^{(n/2)-1} = 1$ from the set of affine functions. If $n = 2$, the numbers of affine and bent functions coincide.

For any even $n \geqslant 4$, a Boolean function $f(x) = x_1 x_2 \oplus x_3 x_4 \oplus \cdots \oplus x_{n-1} x_n$ is a classic example of a bent function. It is easy to prove that it is bent.

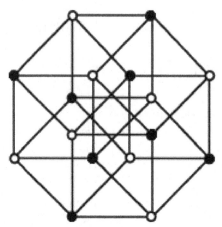

Affine and bent functions in two variables

Indeed, let $x' = (x_1, x_3, \ldots, x_{n-1})$ and $x'' = (x_2, x_4, \ldots, x_n)$ be binary vectors of length $n/2$ with odd and even coordinates of x, respectively. Then $f(x) = f(x', x'') = \langle x', x'' \rangle$ for all x. Consider an arbitrary Walsh-Hadamard coefficient of f; then, $W_f(y) = \sum\limits_{x \in \mathbb{F}_2^n} (-1)^{\langle x, y \rangle \oplus f(x)} = \sum\limits_{x', x'' \in \mathbb{F}_2^{n/2}} (-1)^{\langle x', y' \rangle \oplus \langle x'', y'' \rangle \oplus \langle x', x'' \rangle} =$

$\sum\limits_{x'' \in \mathbb{F}_2^{n/2}} (-1)^{\langle x'', y'' \rangle} \sum\limits_{x'' \in \mathbb{F}_2^{n/2}} (-1)^{\langle x', y' \oplus x'' \rangle} = \sum\limits_{x'' = y'} (-1)^{\langle x'', y'' \rangle} \cdot 2^{n/2} = (-1)^{\langle y', y'' \rangle}$
$\cdot 2^{n/2}$. Hence, f is bent.

Here we used the well-known fact that

$$\sum\limits_{a \in \mathbb{F}_2^k} (-1)^{\langle a, b \rangle} = \begin{cases} 2^k, & \text{if } b \text{ is } 0; \\ 0, & \text{otherwise.} \end{cases}$$

Another simple example is the monomial bent function of Dillon, $f(c) = f(\alpha c^{2^{n/2}-1})$, where c runs through $\mathbb{F}_{2^n}$. There are coefficients $\alpha \in \mathbb{F}_{2^n}$ such that f is bent; for details, see Section 9.4.

2.4 IF n IS ODD?

If n is odd, then everything is completely different. First, the exact upper bound for nonlinearity of a Boolean function in n variables is still unknown! This question is as attractive as it is complicated. Some results on it can be found in articles by Maitra and Sarkar [250] and Kavut et al. [194], among others.

A Boolean function is called *maximal nonlinear* if its nonlinearity is as big as possible. Recall that if n is even, such a definition coincides with the definition of a bent function. In this book, we do not study maximal nonlinear functions for odd n. In what follows, we assume that n is even.

2.5 OPEN PROBLEMS

There are too many open problems in the area of bent functions! Here we briefly discuss some of them.

Extended affine classification. Obtain an extended affine classification of bent functions; at least for the small $n = 10, 12, \ldots$. Necessary definitions can be found in Chapter 5; details of extended affine classification for $n \leqslant 8$ are discussed in Chapter 7.

Equivalent representations. Bent functions can be represented in terms of Hadamard matrices, difference sets, block schemes, linear spreads, sets of subspaces in the Boolean cube, strongly regular graphs, and bent rectangles; see Chapter 6. The problem is to obtain a new equivalent representation of bent functions in order to get a classification of them.

Exact number of bent functions. It is known that there are exactly 8, 896, 5 425 430 528 $\simeq 2^{32.3}$, and $2^9 \times 193\,887\,869\,660\,028\,067\,003\,488\,010\,240 \simeq 2^{106.29}$ bent functions in two, four, six, and eight variables, respectively; see Chapter 7. But what is the exact number of bent functions if $n \geqslant 10$?

Constructions. Propose new direct constructions of bent functions. For now the simplest constructive class of bent functions is the Maiorana-McFarland class; see Chapter 8. Are there other (larger) classes of bent functions with examples that are so easily constructed?

Bent exponents. There are five known distinct bent exponents. A number d is a *bent exponent* if $a \in \mathbb{F}_{2^n}$ such that $f(c) = \text{tr}(ac^d)$ is a bent function; see Chapter 9. Are there other bent exponents for an arbitrary even n? Describe all of them.

General algebraic approach. Propose an algebraic characterization of bent functions—that is, find necessary and sufficient conditions on a trace form (or polynomial form) of a Boolean function to be bent. Think about other algebraic representations of bent functions; see Chapter 9 for details.

Algorithms. Develop effective algorithms for generation of bent functions. Some algorithms are discussed in Chapter 7.

Cryptographic properties. Study the connections between bentness and other cryptographic properties. For example, prove that there are bent functions in n variables of the maximal possible algebraic immunity $n/2$; for this you need to prove the special combinatorial conjecture; see Chapter 10 for details.

Distances between bent functions. Let GB_n be the graph on bent functions in n variables as vertices with edges between functions that are at the minimal possible distance $2^{n/2}$ from each other. It was proven in 2014 that a vertex degree in GB_n is not more than $2^{n/2} \prod_{i=1}^{n/2}(2^i + 1)$. Moreover, only quadratic bent functions have such a degree. Since there are nonweakly normal bent functions if $n \geqslant 14$, this graph is not connected. But is it connected after elimination of all pendant vertices (corresponding to nonweakly normal bent functions)? This and similar open questions are discussed in Chapter 11.

Duality. Let $\mathcal{A}$ be a subset of $\mathbb{F}_2^n$. Let $\mathcal{B}$ be the set of all binary vectors from $\mathbb{F}_2^n$ that are at the maximal possible distance from the set $\mathcal{A}$. Now let $\mathcal{A}'$ be the set of all vectors that are at the maximal possible distance from $\mathcal{B}$. We call a set $\mathcal{A}$ *metrically regular* if $\mathcal{A} = \mathcal{A}'$. In the case of regular sets, it is possible to say that there is *duality* between the definitions of $\mathcal{A}$ and $\mathcal{B}$: the set $\mathcal{A}$ defines $\mathcal{B}$ and vice versa. In Chapter 12, it is proven that the set of vectors of all bent functions (affine functions) is metrically regular. Are there other such sets in the Boolean cube? Give a classification of them.

Lower and upper bounds. Obtain the new (better) upper and lower bounds for the number of bent functions in n variables. For now there is a large gap between the simple lower ($2^{2^{(n/2)+\log_2(n-2)-1}}$) and upper ($2^{2^{n-1}+\frac{1}{2}\binom{n}{n/2}}$) bounds for this number. There are several improvements of these bounds, but they are not too big; see Chapter 13.

Asymptotic value. Find the asymptotic value of the number of all bent functions in n variables. Some details can be found in Chapter 13.

Bent decomposition problem. Is it true that an arbitrary Boolean function in n variables (n is even, $n \geqslant 2$) of degree not more than $n/2$ can be represented as the sum of two bent functions in n variables? If $n \leqslant 6$, the answer is yes. Details are discussed in Chapter 14.

Optimal codes. A code C of length 2^n is a *constant-amplitude code* if every nonzero code word is a vector of values of some bent function in n variables. As we will see in Chapter 4 such codes are very important for transmission in code division multiple access (CDMA; this technology is actively used by most mobile equipment providers). The following natural problem arises: How does one get constructions of linear optimal constant-amplitude codes? What is the maximal possible dimension of such a code for any even n?

Generalizations. Generalizations of bent functions with respect to their algebraic, combinatorial, and cryptographic properties are becoming more numerous and widely studied from year to year. In this book, we consider at least 25 distinct generalizations; see Chapters 15, 16, and 17. The problem is to determine connections between distinct generalizations and propose new constructions of generalized bent functions.

For other open problems in bent functions, we recommend readers consult the papers by Carlet [49], [50]. Several tasks related to open problems in bent functions and their generalizations were stated at the International Student's Olympiad in Cryptography, NSUCRYPTO, in 2014 [11], [286].

2.6 SURVEYS

Surveys of results in bent functions can be found in the following articles and books (we list them in chronological order):

- Technical report "On bent functions" (1966) by Rothaus [317]. It became available in 1976 as [318].
- Technical report "A survey of bent functions" (1972) by Dillon [107].
- Article "A survey of some recent results on bent functions" (2004) by Dobbertin and Leander [112].
- Monograph *Boolean Functions in Coding Theory and Cryptography* (2004 and 2012) by Logachev et al. [243, 247, chapter 6].
- Diploma thesis "Bent functions" (2006) by Neumann [285].
- Article "Monomial bent functions" (2006) by Leander [230].
- Monograph *Cryptographic Boolean Functions and Applications* (2009) by Cusick and Stănică [96, chapter 5].
- Articles "Bent functions: results and applications. A survey of publications" (2009, in Russian) and "Generalizations of bent functions. A survey" (2010) by the author [359, 361].
- Chapters "Boolean functions for cryptography and error-correcting codes" and "Vectorial Boolean functions for cryptography" by Carlet [46, 47] in the monograph *Boolean Models and Methods in Mathematics, Computer Science, and Engineering* (2010).
- Monograph *Nonlinear Boolean Functions: Bent Functions and Their Generalizations* (2011, in Russian) by the author [362].
- Survey "Crosscorrelation of m-sequences, exponential sums, bent functions and Jacobsthal sums" (2011) by Helleseth and Kholosha [161].

- Article "Bent and hyper-bent functions in polynomial form and their link with some exponential sums and Dickson polynomials" (2011) by Mesnager [271].
- Survey "Open problems on binary bent functions" (2013) by Carlet [49].
- Chapter "Bent functions and their connections to combinatorics" (2013) by Helleseth and Kholosha [162] in the monograph *Surveys in Combinatorics*.
- Survey "Open questions on nonlinearity and on APN functions" (2014) by Carlet [50].

The following surveys related to bent functions would also be useful for the interested researcher:

- PhD thesis "Elementary Hadamard difference sets" (1974) of Dillon [108].
- Article "Reed–Muller codes (a survey of publications)" (1996, in Russian) by Kuznetsov and Shkarin [219].
- Article "Codes, bent functions and permutations suitable for DES-like cryptosystems" (1998) by Carlet et al. [51].
- Monograph *Selected Theorems of the Initial Course of Cryptography* (2005, in Russian) by Agibalov [6].
- PhD thesis "Cryptographic properties of Boolean functions and S-boxes" (2006) of Braeken [24].
- Monograph *Combinatorial Properties of Discrete Structures and Applications to Cryptology* (2011, in Russian) by Tarannikov [351].
- Habilitation thesis "Contributions on Boolean functions for symmetric cryptography and error correcting codes" (2012) of Mesnager [272].
- Habilitation thesis "Construction and analysis of cryptographic functions" (2013) of Budaghyan [25].
- PhD thesis "Bent functions affine on subspaces and their metrical properties" (2014, in Russian) of Kolomeec [212].
- Presentation "On perfect and almost perfect nonlinear functions" (2014, in Russian) by Glukhov [138].
- Monograph *Boolean Functions in Cryptography* (2014, in Russian) by Pankratova [295].

CHAPTER 3

History of Bent Functions

INTRODUCTION

Historical aspects of the invention of bent functions are discussed in this chapter. Who was the first mathematician to consider bent functions? This question remains without an exact answer. O. Rothaus was the recognized authority in this area: he introduced bent functions in 1966; his fundamental paper was declassified in 1976 and is well known to everybody who studies bent functions. But not many researchers know that in 1962 V.A. Eliseev and O.P. Stepchenkov proved that the degree of a bent function is not more than $n/2$ if $n \geqslant 4$; they proposed an analog of the McFarland construction 11 years before the publication by R.L. McFarland. They studied bent functions in the Soviet Union, calling them *minimal functions*, and published their results as technical reports that have still not been declassified.

In this chapter, we discuss what is known about this very early study of bent functions and include exclusive photographs of the first researchers. Most of these photographs have never been published before.

3.1 OSCAR ROTHAUS

Oscar Rothaus (1927–2003) was the recognized authority in this area. Bent functions were introduced by him in the 1960s (namely, in 1966).

He graduated from Princeton University. He served in the US Army Signal Corps during the Korean War, and then as a mathematician at the National Security Agency. From 1960 to 1966, he worked at the Defense Department's Institute for Defense Analyses.

He did fundamental research on cryptology, characterized by a top official from the Institute for Defense Analyses follows: "His contributions were considerable. His lectures were elegant and memorable. He was one of the most important teachers of cryptology to mathematicians and mathematics to cryptologists."

Bent Functions
http://dx.doi.org/10.1016/B978-0-12-802318-1.00003-0
25

Oscar Rothaus

In 1966, he wrote a famous paper on bent functions [317]. It was declassified in 1976 [318]. In that article, the main properties of bent functions were obtained, simple constructions of bent functions were given, and several steps for the classification of bent functions in six variables were made. In 1966, he joined Cornell University as a professor and worked there until 2003.

3.2 V.A. ELISEEV AND O.P. STEPCHENKOV

In the USSR, bent functions were also studied in the 1960s. The names of the first Soviet researchers of bent functions are not too public. Also, their papers in this area have still not been declassified.

It is known [137] that Yu. A. Vasiliev, B.M. Kloss, V.A. Eliseev, and O.P. Stepchenkov studied properties of the Walsh-Hadamard transform of a Boolean function at that time. In 1960, they studied the *statistical structure* of a Boolean function—that is, values $\Delta_a^f = 2^{n-1} - \text{dist}(f, \ell_{a,0}) = W_f(a)/2$, where a runs through $\mathbb{F}_2^n$. The notion of a *minimal function* was introduced in the USSR by V.A. Eliseev and O.P. Stepchenkov. A Boolean function is *minimal* if the parameter $\Delta^f = \max_a |\Delta_a^f|$ takes the minimal possible value $2^{(n/2)-1}$. Such functions exist only if n is even. Obviously, "minimal function" is just another name for "bent function." An analog of the McFarland construction of bent functions was proposed by V.A. Eliseev in 1962. He proved that functions of the following form are bent: $f(x_1, \ldots, x_k, \gamma_1, \ldots, \gamma_k) = \sum_{i=1}^{2^k} \delta_i \gamma_j \oplus g(\gamma_1, \ldots, \gamma_k)$, where δ_i are arbitrary conjunctions of variables $\gamma_1, \ldots, \gamma_k$, γ_j are arbitrary linear combinations of variables $x_1, \ldots, x_k$, and g is a Boolean function.

Oleg P. Stepchenkov

Oleg P. Stepchenkov was born in 1936, in Samsonovo, Smolensk Oblast, and graduated from Moscow State University (Department of Mechanics and Mathematics) in 1959. Then he worked at the Penza Research Electrotechnical Institute, where he was Head of the Laboratory for Mathematical Research. In 1982, he received the state prize of the USSR for his great contribution to development of technical means of cryptographic protection for information transmission.

V.A. Eliseev

In 1962, V.A. Eliseev and O.P. Stepchenkov proved that the degree of a bent function is not more than $n/2$ if $n \geq 4$; see [221]. Although this was done 4 years *before* Oscar Rothaus's investigation, it is difficult now to give the credit in bent functions to Soviet mathematicians since their reports have still not seen the light of day.

3.3 FROM THE 1970s TO THE PRESENT

In the early 1970s, bent functions were studied by the American mathematicians J.F. Dillon , P.J. Chase, and K.D. Lerche [82, 107]—in connection with differential sets—and Robert Lee McFarland [260], who constructed the best known large class of bent functions.

One of the best known classical works related to bent functions was written by John Francis Dillon. It was his PhD dissertation entitled "Elementary Hadamard difference sets" [108], which he defended at the University of Maryland, College Park, in 1974.

Since the 1980s, bent functions have widely studied all over the world. There are hundreds of papers about bent functions and related topics. Several constructions of bent functions have been obtained, but the whole class of bent functions in n variables has still not been described. For example, there are no good lower and upper bounds for the number of bent functions. The little progress made in this area is due to the high complexity of the problems, although they can be formulated in a simple manner.

It is interesting that Oscar Rothaus wrote in one of his last annual reports (2001–2002) that he had returned to his earlier interests in coding theory and bent functions and had obtained new, peculiar results [94]. Unfortunately these results remain unpublished.

At the crossroads

CHAPTER 4

Applications of Bent Functions

INTRODUCTION

Distinct applications of bent functions are discussed. First, we consider the use of bent functions in cryptography for constructing the ciphers CAST, Grain, and hash function HAVAL. Further we review connections between bent functions and distinct objects of discrete mathematics and coding theory (Hadamard matrices, strongly regular graphs, and Reed-Muller and Kerdock codes). Some applications to mobile networks are considered (constant-amplitude codes for code division multiple access, CDMA).

4.1 CRYPTOGRAPHY: LINEAR CRYPTANALYSIS AND BOOLEAN FUNCTIONS

Originally, bent functions were introduced in connection with cryptographic applications. Recall that a bent function is a Boolean function in an even number of variables that can be approximated by affine functions in an extremely bad manner. This base property of bent functions is used for protecting ciphers against linear cryptanalysis. Let us say several words about this method.

The linear cryptanalysis method for the fast data encipherment algorithm (FEAL) block cipher was proposed in 1992 by Matsui and Yamagishi [259], and for the Data Encryption Standard (DES) cipher was proposed in 1993 by Matsui [257]. Nowadays, this method (along with the differential cryptanalysis of Biham and Shamir [21]) is reputed to be one of the most efficient statistical methods. The idea of the method is as follows. First, for a known ciphering algorithm, a linear relation L on bits of a plaintext, ciphertext, and key is found that holds with probability $p = 1/2 + \varepsilon$ far enough from $1/2$ (the value ε is called here a *bias*). Then, for a fixed unknown key K, a cryptanalyst collects statistics of N pairs {plaintext—the corresponding ciphertext} and on the basis of the statistics, taking into account the sign of ε, distinguishes between two simple statistical hypotheses: whether or not the relation L holds for this unknown key K. As a result, a new probabilistic relation for bits of

K is found. For this method to work reliably, N should be proportional to $|\varepsilon|^{-2}$.

There are many papers devoted to various generalizations and applications of the linear cryptanalysis method. We describe some of them. A detailed analysis of the linear cryptanalysis method (in particular, for DES) is given by K. Nyberg; see also papers of other authors [29, 98, 151, 258]. To improve the efficiency of the linear cryptanalysis method, Kaliski and Robshaw [191] proposed considering several linear approximations for one combination of key bits; this subject was further developed by Biryukov et al. [22]. In 1997, Sakurai and Furuya [320] presented a way to improve the linear cryptanalysis method (in particular, for the LOKI91 cipher) by considering probabilistic behavior of some bits in approximation instead of their fixed values. See also [14, 203, 283, 324].

A series of papers are devoted to problems of resistance of various ciphering algorithms to linear cryptanalysis. Problems in the construction of Feistel-type ciphering schemes resistant to methods of linear and differential cryptanalysis were considered by Knudsen [202]; for SP networks such problems were studied by Heys and Tavares [164]. In 2001, it was proven by Shorin et al. [326] that the Russian GOST 28147-89 algorithm (with at least five rounds of ciphering for linear cryptanalysis, and seven rounds for differential cryptanalysis) is resistant to these methods. Resistance of the ciphers RC5, RC6, International Data Encryption Algorithm (IDEA), Serpent, Advanced Encryption Standard (AES), Blowfish, and Khufu to the linear cryptanalysis method was analyzed in [20, 23, 152, 253, 282].

In the next section, we illustrate how resistance of a cipher depends on the nonlinearities of the Boolean functions used in its construction.

4.2 CRYPTOGRAPHY: ONE HISTORICAL EXAMPLE

Let us discuss one historical example. In 1993, Matsui [257] showed that DES is not resistant to linear cryptanalysis. For almost 20 years (from 1980 to 1998), DES was a standard of symmetric encryption in the USA. The weakness of the cipher consisted in the "bad" cryptographic properties of its nonlinear components—S-boxes. Mathematically, an S-box is a vectorial Boolean function that maps n input bits to m output bits. In DES, there are eight distinct S-boxes, completely defined in the standard (they can be easily found, for example, on the Internet). Every S-box of DES maps six bits to four bits. On the figure below one can see the round function of DES. Here S-boxes play the *crucial* role in providing the resistance of DES to linear cryptanalysis.

The fifth S-box of DES (denote it by S_5) has very weak nonlinear characteristics. Namely, if $S_5(x_1, x_2, x_3, x_4, x_5, x_6) = (y_1, y_2, y_3, y_4)$, then for 52 inputs (between 64 possible) the following equality holds: $x_2 = y_1 \oplus y_2 \oplus y_3 \oplus y_4 \oplus 1$. In cryptanalytic terms, the maximal bias of S_5 is 20, since $52/64$ differs from $1/2$ by $20/64$. In terms of Boolean functions, this means that a Boolean function $f(x_1, x_2, x_3, x_4, x_5, x_6) = y_1 \oplus y_2 \oplus y_3 \oplus y_4 \oplus 1$ can be easily approximated by the linear function x_2. Indeed, the nonlinearity of f is $N_f = 10$ (f differs from the function x_2 in ten positions). Remember that the maximal possible nonlinearity of a Boolean function in six variables is 28 (see Theorem 14)—an impressive difference! But as we will see in Section 10.5 that it is impossible to construct a S-box mapping six bits to four bits in such a way that all its component functions are bent.

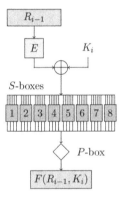

Round function of DES

This weakness of S_5 allowed Matsui [257] to obtain a linear approximation for 14 rounds of DES that holds with a probability of 0.50000057. Then Matsui showed that by using about 2^{43} known plaintext-ciphertext pairs it is possible to guess a DES key (56 bits) with a probability of 0.85. The success of differential cryptanalysis [21] of DES was again caused by weak S-boxes.

Resistance to the types of cryptanalysis mentioned can be provided by using bent functions and their analogs in S-box construction; see [1, 288]. This was done for instance in the ciphers CAST and AES.

Note that application of bent functions (and their vectorial analogs) to S-boxes is not always good. Remember that bent functions are never balanced and correlation immune (for more details, see Chapter 10). Moreover, using them can sometimes decrease the resistance of a cipher.

For such an example, see the paper of Canteaut and Videau [34] on higher-order differential attacks on iterated block ciphers using almost bent round functions.

4.3 CRYPTOGRAPHY: BENT FUNCTIONS IN CAST

The block cipher CAST was introduced in 1997 by the Canadian cryptographers Carlisle Adams and Stafford Tavares [70]. This cipher was approved by the government of Canada as an official replacement for DES. Although CAST-128 is patented by Entrust Technologies, it can be freely used for commercial and noncommercial purposes throughout the world. This cipher is distributed very widely. For instance, CAST-128 is a cipher on default in certain versions of the programs GNU Privacy Guard (GPG) and Pretty Good Privacy (PGP). In addition, CAST-128 is a predecessor of the known cipher CAST-256 (a participant in the AES competition).

This cipher is a classic 16-round Feistel network that encrypts blocks of length 64. The key size is 128 bits. The round function F_i of CAST depends on the number of a round. Below you can see a scheme of it.

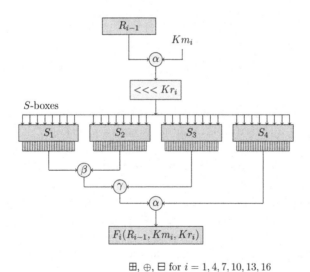

Round function of CAST

There are four S-boxes in the round function, each maps 8 bits to 32 bits. S-boxes are completely defined in the cipher, and can be easily found in the literature. It is natural to represent S_k, the S-box number k, as the collection of 32 Boolean functions $f_j^{(k)}$:

$$S_k(x_1, \ldots, x_8) = (y_1, \ldots, y_{32}), \quad \text{where}$$
$$y_j = f_j^{(k)}(x_1, \ldots, x_8), \ j = 1, \ldots, 32.$$

CAST is designed in such a way that all functions $f_j^{(k)}$ are bent. Moreover, any linear combination of bent functions $f_j^{(k)}$ for a fixed k has high nonlinear properties. More details on choosing bent functions for CAST can be found in [2].

4.4 CRYPTOGRAPHY: BENT FUNCTIONS IN GRAIN

Grain is a stream cipher from the eSTREAM project. It was designed by Hell et al. [155]. It has been selected for the final eSTREAM portfolio (profile 2). Grain is designed primarily for restricted hardware environments. It accepts an 80-bit key and a 64-bit initial vector. The cipher consists of a nonlinear feedback shift register (NFSR) and a linear feedback shift register, both of length 80 bits.

The nonlinear feedback polynomial of the NFSR, $g(x)$, is constructed as the sum of a linear function and a bent function. Such a combination is suggested by Hell et al. [155] in order to provide high resiliency and nonlinearity. The simplest quadratic bent function

$$b(x) = x_0 x_1 \oplus x_2 x_3 \oplus x_4 x_5 \oplus x_6 x_7 \oplus x_8 x_9 \oplus x_{10} x_{11} \oplus x_{12} x_{13}$$

in 14 variables was taken. Its nonlinearity equals 8128. To increase the resiliency, five linear terms were added to the function. This results in a balanced function with a resiliency of 4 and nonlinearity of $2^5 \times 8128 = 260\,096$.

Note that if two arbitrary Boolean functions ξ and ψ in $n+1$ and n variables, respectively, are connected by the equality $\xi(x, x_{n+1}) = \psi(x) \oplus x_{n+1}$ for all $x \in \mathbb{F}_2^n$, $x_{n+1} \in \mathbb{F}_2$, then their nonlinearities are linked as $N_\xi = 2 \cdot N_\psi$.

The resulting Boolean function for the NFSR of Grain has 19 variables and is cheap to implement for hardware. The best linear approximation gives a linear function in 19 variables that contains at least all the linear terms of the function. There are 2^{14} such functions, and they have bias $\varepsilon_g = 2^{-8}$.

It is interesting that the filter function h in five variables that transforms four bits from the linear feedback shift register and one bit from the NFSR to a single bit was not designed very carefully by the authors of the cipher. Namely, the function

$$h(x) = x_1 \oplus x_4 \oplus x_0 x_3 \oplus x_2 x_3 \oplus x_3 x_4 \oplus x_0 x_1 x_2 \oplus x_0 x_2 x_3$$
$$\oplus x_0 x_2 x_4 \oplus x_1 x_2 x_4 \oplus x_2 x_3 x_4$$

is equal to zero when the arguments x_1, x_3, and x_4 are equal to zero independently of the values of x_0 and x_2. In other words, the Boolean function h is a k-normal Boolean function, where $k = 2$. In 2011, Mihaljevic et al. [276] proposed a cryptanalysis of Grain based on this weakness of the Boolean function h.

4.5 CRYPTOGRAPHY: BENT FUNCTIONS IN HAVAL

The hash function HAVAL was designed by Zheng et al. [402] in 1993. An updating algorithm is the main part of it. It is a vectorial Boolean function that maps $256 + 1024$ bits to 256 bits. The updating algorithm H processes a block in three, four, or five passes, which is specified by the three–bit field PASS in the last block. Each of the five passes H_1, H_2, H_3, H_4, and H_5 has 32 rounds of operations. In constructions of H_i distinct Boolean functions f_i in seven variables are used. These Boolean functions are of central importance to the hashing algorithm. Note that they were obtained from four (all) affinely nonequivalent bent functions in six variables:

$$g_1(x_6, \ldots, x_1) = x_1 x_4 \oplus x_2 x_5 \oplus x_3 x_6;$$

$$g_2(x_6, \ldots, x_1) = x_1 x_2 x_3 \oplus x_2 x_4 x_5 \oplus x_1 x_2 \oplus x_1 x_4 \oplus x_2 x_6 \oplus x_3 x_5 \oplus x_4 x_5;$$

$$g_3(x_6, \ldots, x_1) = x_1 x_2 x_3 \oplus x_1 x_4 \oplus x_2 x_5 \oplus x_3 x_6;$$

$$g_4(x_6, \ldots, x_1) = x_1 x_2 x_3 \oplus x_2 x_4 x_5 \oplus x_3 x_4 x_6 \oplus x_1 x_4 \oplus x_2 x_6 \oplus x_3 x_4 \oplus x_3 x_5$$

$$\oplus x_3 x_6 \oplus x_4 x_5 \oplus x_4 x_6.$$

Namely, functions f_i, where $i = 1, 2, 3, 4$, are obtained from g_i as

$$f_i(x_6, \ldots, x_0) = g_i(x_6, \ldots, x_1) \oplus x_0 x_i \oplus x_0.$$

The fifth function f_5 is given by

$$f_5(x_6, \ldots, x_0) = g_1(x_6, \ldots, x_1) \oplus x_0 x_1 x_2 x_3 \oplus x_0 x_5 \oplus x_0.$$

Boolean functions g_i have nonlinearity equal to $2^6 - 2^3 = 56$, which is the maximum possible for Boolean functions in seven variables. All these Boolean functions are linearly inequivalent.

The properties of functions f_i mentioned are important for making HAVAL resistant to linear attacks.

4.6 HADAMARD MATRICES AND GRAPHS

The classical combinatorial problem of constructing *Hadamard matrices* is well known and remains unsolved since 1893. In the case, when the size of the matrix is $N = 2^n$ (n is even), this problem can be transformed (with some restrictions) to the task of constructing bent functions in n variables [318] (see Theorem 24).

In finite field theory, maximal nonlinear Boolean functions are connected to the *elementary Hadamard difference sets* of the special type (see [247] and Theorem 26).

Bent functions are closely connected to strongly regular graphs. Recall that a graph is called *strongly regular with parameters* (v, k, λ, μ) if it contains v vertices each of degree k and for any vertices x and y the number of vertices incident to x and y simultaneously is equal to λ or μ and it depends on the presence or absence of the edge between x and y. It was proven in [17] that a Boolean function f is bent if and only if its Cayley graph G_f is strongly regular and $\lambda = \mu$; see Theorem 30.

Following the results in [17], Huang and You [177] proposed feasible parameters and the corresponding eigenvalues of the associated Cayley graphs of bent functions. In particular, they included all of the graphs with at most 280 vertices.

In 2010, Tan et al. [348] proved a new characterization of weakly regular ternary bent functions via partial difference sets (that correspond to strongly regular graphs as is known). Using known families of bent functions, they obtained new families of strongly regular graphs, some of which were previously unknown. Moreover, they gave a new proof that the Coulter-Matthews and ternary quadratic bent functions are weakly regular [348]. Generalization of this technique from the ternary case to an arbitrary p-ary case was proposed in 2011 by Chee et al. [84].

4.7 LINKS TO CODING THEORY

In coding theory, there is a well-known task of determining the covering radius for the *Reed-Muller code* $\mathrm{RM}(\ell, n)$. This task is equivalent (if the code has order 1) to the task of finding the most nonlinear Boolean functions [194, 250].

Special sets of quadratic bent functions allow one to construct *Kerdock codes* [195] that are optimal and have large code distances that grow with the code lengths [102, 328]. This very optimality of Kerdock codes is caused by extremal properties of bent functions.

One of the definitions of Kerdock codes is as follows: Let functions $g_1, \ldots, g_{2^{n-1}-1}$ be distinct quadratic bent functions in n variables (n is even, $n \geqslant 6$) such that the sum of any two of them is again a bent function. Let $\mathcal{K}_n$ be the following set of Boolean functions: it is a union of the class $\mathcal{A}_n$ of all affine Boolean functions in n variables and its $2^{n-1} - 1$ cosets—namely, $\mathcal{A}_n \oplus g_i$, where $i = 1, \ldots, 2^{n-1} - 1$. Note that $|\mathcal{K}_n| = 2^{n+1} \cdot 2^{n-1} = 2^{2n}$ since cosets do not intersect. A *Kerdock code* of length 2^n consists of vectors of values for all functions from $\mathcal{K}_n$. The code distance of the Kerdock code is big enough: it is equal to 2^{n-1}.

So, to construct a Kerdock code one should construct such a special system of bent functions $\{g_i\}$, $i = 1, \ldots, 2^{n-1} - 1$, in fact not necessarily quadratic. Note that this task can be transformed to the task of finding *orthogonal spreads* in the finite vectorial space [193], and it is a very good example of the connection between bent functions and extremal geometrical objects.

So-called *Kerdock sets* (or *Kerdock ensembles*) are even better known in information theory for their special properties. In 1989 (the publication in English was in 1991), it was proven by Nechaev [284] that Kerdock codes can be cyclically closed—this fact simplified the method of constructing Kerdock codes (and based on them Kerdock ensembles), and as a result made them attractive for applications in navigation systems (Global Navigation Satellite System) such as GLONASS and GPS.

As another example from the coding theory, we mention so-called *bent codes*—linear binary codes with columns of parity-check matrices taken from the supports of bent functions [46]. Note that these bent codes have the maximal possible code dimensions.

Bent functions are intensively used in constructing codes for code division multiple access (CDMA) systems. We discuss this topic in Section 4.9.

As for new applications to coding, we mention the following.

In 2013, Kim and Jung [200] proposed a new forward error correction method using a coset constructed from a bent function and a Reed-Muller code for providing accurate dimming control in on-off keying-based *visible light communication systems*.

In 2012, Carlet et al. [64] proposed a class of bent functions on a Galois ring. On the basis of this class, systematic *authentication codes* were then presented. Later, Ku-Cauich and Tapia-Recillas [217] used a class of bent functions on a Galois ring of characteristic p^2 and the Gray map on this ring to construct new classes of systematic authentication codes. Connections with authentication schemes can be found in [55].

For some relations between bent functions and *cyclic codes*, see Wolfmann [379, 380].

4.8 BENT SEQUENCES

Sets of *bent sequences* with elements $+1$ and -1 constructed using bent functions have the lowest values of mutual correlations and autocorrelations (reach the Welch bound). That is why these sets of bent sequences are used successfully in the communication systems of multiple access.

The first and the best known work on bent sequences is that of Olsen et al. [294], published in 1982.

Here we collect several links on bent sequences and invite the interested reader to follow them.

In 1987, Losev [248] studied aspects of decoding of the bent function sequences by fast Hadamard transform. In 1990, Adams and Tavares [4] proposed methods for generating and counting binary bent sequences. In 1993, Matsufuji and Imamura [255] investigated balanced quadriphase sequences with optimal periodic correlation properties constructed with real-valued bent functions (i.e., generalized bent functions with $q = 4$; for these functions, see also Section 15.6). In 1995, Gabidulin [126] considered a partial classification of sequences with perfect autocorrelation and bent functions. In 1998, Ipatov and Kamaletdinov [182] studied groups of periodic bent function sequences.

In 2002, Paterson [299] analyzed applications of bent sequences for orthogonal frequency-division multiplexing (OFDM) and multicode CDMA, and studied related problems in algebraic coding theory.

Shannon's lecture at the cryptographic conference, twenty-first century

Applications of quadratic bent functions to bent sequences were considered by Yu and Gong [391]. Further constructions and generalizations of bent function sequences were studied by Xia et al. [383] in 2009.

4.9 MOBILE NETWORKS, CDMA

Bent functions are used in CDMA. The CDMA technology for digital mobile services was standardized in 1993 by the US Telecommunication Industry Association as the standard IS-95 (Mobile Station–Base Station Compatibility Standard for Dual-Mode Wideband Spread Spectrum Cellular System). Presently, the technology is actively used by most mobile equipment providers throughout the world in accordance with the third-generation mobile service standard IMT-2000 (in Russia, the standard IMT-MC 450 or CDMA-450). Note that the first article devoted to this technology was published in the USSR as long ago as 1935 by Ageev [5]. The CDMA systems use broadband signals, and many clients simultaneously use the whole band of frequencies of the channel. Since every client is assigned a unique code, it is easy to isolate this code from the "noise." CDMA systems substantially increase the bandwidth of the channel and are quite efficient.

In 2000, Wada [369] established a connection between bent functions and codes for CDMA (see also the article by Paterson [300]). Consider the simplest model of information transmission in a multicode CDMA system.

For $N = 2^n$, take a size $N \times N$ Hadamard matrix $A_N = (a_{jt})$ of Sylvester type. There are N parallel data flows. We can represent the transmitted information as a binary vector c of length N (one bit from each flow). The signal in multicode CDMA is modeled as

$$S_c(t) = \sum_{j=0}^{N-1} (-1)^{c_j} a_{jt},$$

where $t = 0, 1, \ldots, N - 1$ is a discrete time parameter—that is, the jth row of the matrix A is multiplied by $(-1)^{c_j}$, and the transmitted signal S_c is the sum of these new rows. At every moment of time, one bit of the sequence S_c is transmitted. An important parameter is the *peak to average power ratio* of the signal, which is defined as

$$\text{PAPR}(c) = \frac{1}{N} \max_t |S_c(t)|^2.$$

Note that $1 \leqslant \text{PAPR}(c) \leqslant N$. The quantity $|S_c(t)|^2$ is proportional to the power necessary to transmit this signal; thus, the vectors c with minimal $\text{PAPR}(c)$ are most suitable for transmission. We may assume that the vectors c are chosen from some binary code C of length N. Put $\text{PAPR}(C) = \max_{c \in C} \text{PAPR}(c)$. If $\text{PAPR}(C) = 1$, then C is called a *constant-amplitude code*. Currently, it is a problem to construct a code of this type with large size and large code distance. The following theorem holds [300, 369]:

Theorem 15. *A code C of length 2^n is a constant-amplitude code if and only if every code word is a vector of values of some bent function in n variables.*

Sometimes a zero vector also can be added to a constant-amplitude code. Indeed, given the vector c of values of a Boolean function f in n variables, $\text{PAPR}(c) = \frac{1}{2^n} \max_{x \in \mathbb{F}_2^n} |W_f(x)|^2$. Therefore, bent functions play a substantial role in constructing codes for CDMA systems.

In 2007, Zhou et al. [405] considered codes with low peak-to-average power ratio for multicode CDMA. On the basis of the concept of quarter bent functions, a new inequality relating the minimum order of terms of a bent function and the maximum Walsh spectral magnitude was proven, and it facilitates the generalization of some known results. In particular, a new simple proof of the nonexistence of the homogeneous bent functions of degree n in $2n$ variables for $n > 3$ was obtained without invoking results from the difference set theory. Zhou et al. [405] proposed a new coding approach to achieve the constant-amplitude transmission of code-word length 2^n for both even and odd n.

A technique for constructing constant-amplitude codes of small lengths was proposed by Pavlov [301] in 2010.

In 2008, Bey and Kyureghyan [19] obtained results connected to construction of constant-amplitude codes. Namely, they showed that any *bent set* (i.e., a set of Boolean functions such that the sum of any two distinct functions is a bent function) yields a homogeneous system of linked symmetric designs with the same design parameters as those systems derived from Kerdock sets. They observed that there are bent sets of size equal to the square root of the Kerdock set size which consist of Boolean functions with arbitrary degrees. See also [18].

Constructions of special codes for CDMA from Hadamard matrices and almost bent functions can be found in the articles by Smith et al. [330] and Hunt and Smith [179]. See also the article by Gong [142] related to special signal sets from bent functions. Spreading codes for CDMA constructed from semibent functions were considered in 2012 by Hunt and Smith [178].

Several ideas connected to GLONASS CDMA applications see in paper by Ipatov and Shebshayevich [183].

One algebraic generalization of bent functions related to CDMA is considered in Section 15.6.

4.10 REMARKS

In 2013, Stoyanov and Kordov proposed [344] a modified encryption scheme based on a 256-bit bent Boolean function and feedback with carry shift register. They estimated the properties of output bits by the NIST, DIEHARD, and ENT test packages. On the basis of the results of the cryptanalysis, Stoyanov and Kordov [344] declared that the new cryptographic scheme provides an exclusive level of data security.

Note that some correlations between the internal construction and the properties of bent functions in cryptographic systems were considered by Zhang [396] in 2009.

The generation of pseudorandom sequences from cellular automata and bent functions was discussed by Garcia et al. [135] in 2007.

CHAPTER 5

Properties of Bent Functions

INTRODUCTION

Some basic properties of bent functions are discussed in this chapter. The first one is a restriction on the degree of a bent function: if f is bent in n variables, then $2 \leqslant \deg(f) \leqslant n/2$. The second one is that a Boolean function extended affinely equivalent to a bent function is bent too. Dual bent functions are defined; relations between degrees and algebraic normal form coefficients of bent functions and their dual functions are presented. It is mentioned that a bent function is a nondegenerate function.

5.1 DEGREE OF A BENT FUNCTION

In what follows let n be an even number. The following property of bent functions is very important and plays a key role in proving many results in bent functions [318]. According to Kuz'min et al. [221], it was established as long ago as 1962 by V.A. Eliseev and O.P. Stepchenkov.

Theorem 16. *The degree* $\deg(f)$ *of a bent function* f *in* $n \geqslant 4$ *variables is not more than* $n/2$. *If* $n = 2$, *a bent function is quadratic.*

One can find a simple proof of this fact in the book by Cusick and Stănică [96].

Obviously, a Boolean function of degree less than or equal to 1 cannot be bent. It is easy to see that there are bent functions of all other possible degrees from 2 to $n/2$ if $n \geqslant 4$ (just use the Maiorana-McFarland construction for this; see Theorem 34). For example, the quadratic Boolean function $f(x_1, \ldots, x_n) = x_1 x_2 \oplus x_3 x_4 \oplus \cdots \oplus x_{n-1} x_n$ is bent for any even n.

Note that in 2004 Hou [171] determined the bound for p-ary bent functions—namely, he proved that if f is a p-ary bent function (p is prime) in n variables, then $\deg(f) \leqslant \frac{(p-1)n}{2} + 1$. In addition, if f is weakly regular, then $\deg(f) \leqslant \frac{(p-1)n}{2}$. Necessary definitions and details can be found in Sections 15.2 and 15.3.

Bent Functions
http://dx.doi.org/10.1016/B978-0-12-802318-1.00005-4

5.2 AFFINE TRANSFORMATIONS OF BENT FUNCTIONS

Recall that Boolean functions f and g in n variables are *affinely equivalent* if there is a nondegenerate affine transformation of variables that maps one Boolean function to another. In other words, f and g are affinely equivalent if there is a nonsingular $n \times n$ matrix A and a vector b of length n, such that $g(x) = f(Ax \oplus b)$ for every $x \in \mathbb{F}_2^n$.

Boolean functions f and g in n variables are *extended affine equivalent* (or EA-equivalent) if there is a nondegenerate affine transformation of variables that maps one Boolean function to another up to the addition of an affine function. So, f and g are extended affinely equivalent if there is a nonsingular $n \times n$ matrix A, vectors b and c of length n, and a constant $\lambda \in \mathbb{F}_2$, such that

$$g(x) = f(Ax \oplus b) \oplus \langle c, x \rangle \oplus \lambda \quad \text{for every } x \in \mathbb{F}_2^n.$$

Very often in the literature, equivalence of this type is also called "affine equivalence."

Recall that in what follows by *equivalent Boolean functions* we mean functions that are extended affinely equivalent unless stated otherwise.

Note that the degree of a function is invariant under affine and extended affine equivalences.

It is easy to prove the following very important fact:

Theorem 17. *Let f be a bent function in n variables. Then*

(1) a Boolean function $f(Ax \oplus b)$ is bent, where A is an $n \times n$ invertible matrix over $\mathbb{F}_2$ and b is an arbitrary vector of length n;

(2) a function $f \oplus \ell$ is bent for any affine function ℓ.

Thus, class $\mathcal{B}_n$ is closed up to any nondegenerate affine transformation of variables and addition of any affine Boolean function. In other words, every Boolean function extended affinely equivalent to a bent function is a bent function.

Since the class $\mathcal{B}_n$ is closed up to extended affine equivalence, there arises the problem of extended affine classification of bent functions.

According to Chase et al. [82] (see also Dillon [107]), the following theorem holds:

Theorem 18. *Any quadratic bent function in n variables is extended affinely equivalent to the function $x_1 x_2 \oplus x_3 x_4 \oplus \cdots \oplus x_{n-1} x_n$.*

For bent functions of degree more than or equal to 3, the problem of extended affinely classification is still open. Classifications obtained for small n can be found in Chapter 7. Problems of equivalence of Boolean functions are discussed in more details by Cheremushkin [87].

Item (2) in Theorem 17 invites us to study the following natural question: Is there a nonaffine Boolean function that can be added to an arbitrary bent function and "save" its bent property? In 2010, the answer *no* was given. Namely, in [360], the following theorem was proven:

Theorem 19. *Let f be a Boolean function in n variables of degree $\geqslant 2$. Then there is a bent function g in n variables such that $f \oplus g$ is not bent.*

We prove this theorem in Chapter 12 when study the group of automorphisms of bent functions.

5.3 RANK OF A BENT FUNCTION

Let us give some necessary notions. A *design* (or *block scheme*) with parameters (v, k, λ) is a system of k-element subsets (*blocks*) of a v-element set such that any pair of distinct elements is contained in exactly λ blocks. A design is *symmetric* if the number of blocks is equal to the number of elements. It is known (see Theorem 27) that a Boolean function f in n variables is bent if and only if the system of sets $D_z = D \oplus z$, where $z \in \mathbb{F}_2^{n+1}$ and $D = \{(x, f(x)) | x \in \mathbb{F}_2^n\}$, is a symmetric design with parameters $(2^{n+1}, 2^n, 2^{n-1})$.

Then the rank of a bent function can be defined as follows: the *rank* of a bent function f is the rank of the incidence matrix of the corresponding design. Recall that the *incidence matrix* $A = (a_{ij})$ of the symmetric design is a $v \times v$ binary matrix whose rows and columns are indexed by blocks and elements, respectively, and whose entry a_{ij} is 1 if element j belongs to the ith block and is 0 otherwise.

In 2007, Weng et al. [377] showed that the rank of a bent function is invariant under equivalence. They proposed some upper and lower bounds on ranks of bent functions (in the general case), Maiorana McFarland bent functions, and partial spread bent functions. As a consequence, it was proven in [377] that almost every Desarguesian partial spread bent function is not equivalent to any Maiorana-McFarland bent function.

Later, in 2008, Weng et al. [376] considered ranks of Maiorana-McFarland bent functions. The upper and lower bounds on ranks were given; bent functions that achieve these bounds on rank were determined. As a consequence, nonequivalence of some bent functions was derived.

5.4 DUAL BENT FUNCTIONS

For a bent function f, the *dual function* $\tilde{f}$ in n variables is defined by the equality

$$W_f(y) = 2^{n/2}(-1)^{\tilde{f}(y)}.$$

This definition is correct since $W_f(y) = \pm 2^{n/2}$ for any vector y. It is not hard to prove that the function $\tilde{f}$ is a bent function too. It holds that $\tilde{\tilde{f}} = f$.

Note that if $\deg(f) = n/2$, then $\deg(\tilde{f}) = n/2$. In general, there is the following connection between the degrees of a bent function and its dual [46]:

Theorem 20. *Let f be an arbitrary bent function in n variables. Then,*

$$n/2 - \deg(f) \geqslant \frac{n/2 - \deg(\tilde{f})}{\deg(\tilde{f}) - 1}.$$

The following fact about algebraic normal form (ANF) coefficients of a bent function and its dual is also very important in proving the results concerning bent functions. Its proof can be found in [96, lemma 5.17].

Theorem 21. *Let f be a bent function in n variables, $n \geqslant 4$. Then,*

$$\sum_{x \preccurlyeq y} f(x) = 2^{\mathrm{wt}(y)-1} - 2^{(n/2)-1} + 2^{\mathrm{wt}(y)-n/2} \sum_{x \preccurlyeq y \oplus 1} \tilde{f}(x).$$

In 2013, Çeşmelioğlu et al. [73] studied the problem of "bentness" for functions dual to p-ary bent functions; see Section 15.2. The properties of dual functions for some specially constructed bent functions were studied by several authors. Bent functions dual to Niho functions were studied by Budaghyan et al. [26] and Carlet et al. [62], and bent functions dual to Kasami bent functions were investigated by Langevin and Leander [226].

Self-dual bent functions (such that $f = \tilde{f}$) and anti-self-dual bent functions (such that $f = \tilde{f} \oplus 1$) will be discussed in Section 16.5.

5.5 OTHER PROPERTIES

From the definition of a bent function, the following theorem easily follows:

Theorem 22. *Every bent function in n variables is of Hamming weight $2^{n-1} \pm 2^{(n/2)-1}$.*

According to Theorem 17, if f is bent, then $f \oplus 1$ is also bent. That is why all bent functions can be divided into halves: in one part, there are bent functions of weight $2^{n-1} - 2^{(n/2)-1}$, and in the second part, there are their negations of weight $2^{n-1} + 2^{(n/2)-1}$.

A Boolean function f in n variables *has a degenerate (fictitious) variable x_i* if for any vector $b \in \mathbb{F}_2^n$ it holds that $f(b) = f(b \oplus e_i)$, where e_i is a vector of weight 1 with the ith coordinate being nonzero. In other words, a variable is fictitious if it does not occur in the ANF of f. A Boolean function is *nondegenerate* if it has no fictitious variables.

Theorem 23. *A bent function in n variables is nondegenerate—that is, all variables are presented in its ANF.*

Let us mention some papers related to the properties of bent functions.

In 2000, Hou [170] proved a 2-adic inequality for the coefficients of binary bent functions in their polynomial representations.

In 2012, Glukhov and Zakrevskii [139] introduced and studied the coefficients of additivity and affinity of functions on finite groups. In particular, they considered applications of them to bent functions.

Some other properties can be found in papers by Zhang and Lü [392], by Zhang et al. [395].

Moscow, USSR, in the 1960s

CHAPTER 6

Equivalent Representations of Bent Functions

INTRODUCTION

Several attempts to find for bent functions a new combinatorial or algebraic equivalent representation are considered. As usual, such a new representation shows bent functions from another perspective and can be helpful in describing their properties or counting them. Connections of bent functions to Hadamard matrices, difference sets, block schemes, linear spreads, sets of subspaces in the Boolean cube, strongly regular graphs, and bent rectangles are considered.

6.1 HADAMARD MATRICES

A *Hadamard matrix* is a square $k \times k$ matrix A with elements ± 1 such that $AA^{\mathrm{T}} = kE$, where E is the identity matrix. Let us enumerate the rows and columns of a $2^n \times 2^n$ matrix with binary vectors x and y of length n. Rothaus [318] proved the following theorem:

Theorem 24. *The following statements are equivalent:*

(1) *A Boolean function f in n variables is bent.*

(2) *$A = (a_{x,y})$, where $a_{x,y} = 2^{-n/2} W_f(x \oplus y)$, is a Hadamard matrix.*

(3) *$D = (d_{x,y})$, where $d_{x,y} = (-1)^{f(x \oplus y)}$, is a Hadamard matrix.*

Recall that a Boolean function $D_y f(x) = f(x) \oplus f(x \oplus y)$ is called a *derivative of f with respect to the vector y, where $y \in \mathbb{F}_2^n$. The following theorem is well known as the *propagation criterion PC(n) of order n*. In practice, it gives a very useful equivalent definition of a bent function.

Theorem 25. *A Boolean function f in n variables is bent if and only if for any nonzero vector y its derivative $D_y f(x) = f(x) \oplus f(x \oplus y)$ is balanced—that is, it takes values 0 and 1 equally often.*

6.2 DIFFERENCE SETS

From the beginning, bent functions were studied in connection with difference sets [107]. Let a finite Abelian group G have order v and be

Bent Functions
http://dx.doi.org/10.1016/B978-0-12-802318-1.00006-6

presented in the additive form. A subset $D \subseteq G$ of size k is called *a difference set* with parameters (v, k, λ) if every nonzero element $g \in G$ can be represented in the form $g = b - d$ exactly in λ ways, where b and d are elements of the set D. The following theorem holds [107]:

Theorem 26. *A Boolean function f in n variables is a bent function if and only if the set $D = \{(x, f(x)) | x \in \mathbb{F}_2^n\}$ is a difference set with parameters $(2^{n+1}, 2^n, 2^{n-1})$ in the additive group $\mathbb{Z}_2^{n+1}$.*

Indeed this fact can be easily obtained from Theorem 25. Difference sets with the parameters mentioned in Theorem 26 are called *elementary Hadamard sets*. Examples of such sets were known before bent functions were introduced [107].

6.3 DESIGNS

It is known [149] that difference sets are closely connected to designs (block schemes). Recall that a *design* (or *block scheme*) with parameters (v, k, λ) is a system of k-element subsets (*blocks*) of a v-element set such that any pair of distinct elements is contained in exactly λ blocks. A block scheme is *symmetric* if the number of blocks is equal to the number of elements (i.e., $= v$). The *incidence matrix* $A = (a_{ij})$ of the symmetric design is a $v \times v$ binary matrix whose rows and columns are indexed by blocks and elements, respectively, and whose entry a_{ij} is 1 if element j belongs to the ith block and is 0 otherwise.

Theorem 26 then takes the following form.

Theorem 27. *A Boolean function f in n variables is a bent function if and only if the system of sets $D_z = D \oplus z$, where z runs through $\mathbb{F}_2^{n+1}$ is a symmetric design with parameters $(2^{n+1}, 2^n, 2^{n-1})$. Recall that D is defined as $D = \{(x, f(x)) | x \in \mathbb{F}_2^n\}$.*

Recall that the *rank* of a bent function f is the rank of the incidence matrix of the corresponding design.

In 2007 and 2008, Weng et al. [376, 377] studied ranks of Maiorana-McFarland bent functions, partial spreads, etc.; for more details, see Section 5.3.

6.4 LINEAR SPREADS

In 1997, Yashchenko [386] proposed the following representation of the class of bent functions. The foundation of it is the fact that any Boolean function f in n variables can be represented as the *linear spread*:

$$f(x', x'') = \langle x', h(x'')\rangle \oplus g(x''), \quad \text{where } x' \in \mathbb{F}_2^r,\ x'' \in \mathbb{F}_2^k \qquad (6.1)$$

for the appropriate numbers r and k such that $n = r + k$, mapping $h : \mathbb{F}_2^k \to \mathbb{F}_2^r$ and Boolean function g in k variables. The maximal possible value of r in this representation is called the *linearity index* of a Boolean function f.

A subset M of the space $\mathbb{F}_2^n$ is called a *bent set* if its size is $2^{2\ell}$ for some ℓ, and for any nonzero vector $z \in \mathbb{F}_2^n$ the set $M \cap (z \oplus M)$ has even cardinality or is empty.

A pair $(g; M)$, where g is a Boolean function in k variables and M is a bent set, is called a *partial bent function* if for any $y' \in \mathbb{F}_2^r$ and nonzero $y'' \in \mathbb{F}_2^k$ the function $g(x'') \oplus g(x'' \oplus y'')$ is balanced on the set $M \cap \big((y', y'') \oplus M\big)$.

Theorem 28. *A Boolean function f of the form (6.1) is bent if and only if $n > 2r$ and for any vector $x' \in \mathbb{F}_2^r$ the following hold:*

(1) The size of the set $h^{-1}(x')$ is equal to 2^{n-2r}.

(2) Set $h^{-1}(x')$ is a bent set.

(3) The pair $(g; h^{-1}(x'))$ is a partial bent function.

In 2004, Carlet [41] proposed (independently) the construction of bent functions that is a partial case of the representation considered (see Theorem 37).

6.5 SETS OF SUBSPACES

Let us consider a geometrical representation for the class of bent functions introduced by Carlet and Guillot [61] in 1998 (see also an earlier paper [60]).

Let f be a Boolean function in n variables. Let $\mathrm{Ind}_S : \mathbb{F}_2^n \to \mathbb{F}_2$ be the *characteristic function* of the subset $S \subseteq \mathbb{F}_2^n$—that is, Ind_S takes the value 1 on elements from S and the value 0 on all other elements.

Theorem 29. *A function f is bent if and only if there are subspaces $E_1, \ldots, E_k$ of dimension $n/2$ or $(n/2) + 1$ in the space $\mathbb{F}_2^n$ and nonzero numbers $m_1, \ldots, m_k$ such that for any $x \in \mathbb{F}_2^n$*

$$\sum_{i=1}^{k} m_i \mathrm{Ind}_{E_i}(x) = \pm 2^{(n/2)-1} \mathrm{Ind}_{\{0\}}(x) + f(x).$$

Carlet and Guillot [61] proposed restrictions on the choice of subspaces $E_1, \ldots, E_k$, such that this choice becomes unique for every bent function.

Thus, it is possible to make this representation unique for every bent function.

6.6 STRONGLY REGULAR GRAPHS

Another approach to classification of bent functions was proposed in 1999 by Bernasconi and Codenotti [16]; later they wrote a paper together with J.M. Van der Kam [17].

Consider the *Cayley graph* $G_f = G(\mathbb{F}_2^n, \text{supp}(f))$ of a Boolean function f. All vectors of length n are vertices of the graph. There is an edge between two vertices x and y if vector $x \oplus y$ belongs to $\text{supp}(f)$.

A regular graph G is called *strongly regular* if there are nonnegative integers λ and μ such that for any vertices x and y the number of vertices incident to x and y is both equal to λ or μ and depends on the presence or absence of the edge between x and y. In [17], the following is proven:

Theorem 30. *A Boolean function f is bent if and only if the Cayley graph G_f is strongly regular and $\lambda = \mu$.*

Following the results obtained by Bernasconi et al. [17], Huang and You [177] proposed feasible parameters and the corresponding eigenvalues of the associated Cayley graphs of bent functions. In particular, they included all of the graphs with at most 280 vertices.

In 2010, Tan et al. [348] proved a new characterization of weakly regular ternary bent functions via partial difference sets (that correspond to strongly regular graphs as is known). Using known families of bent functions, they obtained new families of strongly regular graphs, some of which were previously unknown. Moreover, they gave a new proof that the Coulter-Matthews and ternary quadratic bent functions are weakly regular. Generalization of this technique from the ternary case to an arbitrary p-ary case was proposed in 2011 by Chee et al. [84].

6.7 BENT RECTANGLES

In 2000, Agievich [7] found a bijection between the set of all bent functions in n variables and the set of *bent rectangles*.

Let f be a Boolean function in n variables, $n = r + k$. We call the vector W_f of Walsh-Hadamard coefficients of f the *spectral vector* of the function f. Let us represent the vector of values f of a function f in the form $f = (f_{(1)}, \ldots, f_{(2^r)})$, where every vector $f_{(i)}$ has length 2^k. Let $f_{(i)}$ be a

Boolean function in k variables for which $\boldsymbol{f}_{(i)}$ is the vector of values, $i = 1, \ldots, 2^r$. Let us consider the matrix $\mathcal{M}_f$ of size $2^r \times 2^k$ with spectral vectors $\boldsymbol{W}_{f_{(1)}}, \ldots, \boldsymbol{W}_{f_{(2^r)}}$ as rows.

A matrix of size $2^r \times 2^k$ is called a *bent rectangle* if every line (row or column) multiplied by $2^{r-(n/2)}$ is a spectral vector of the appropriate Boolean function. According to [7], the following theorem holds:

Theorem 31. *A Boolean function f is a bent function if and only if the matrix $\mathcal{M}_f$ is a bent rectangle.*

This approach allowed Agievich [7] to find a description of all bent functions in six variables (see later) and obtain an algorithm for constructing the special class of bent functions in n variables for any n (Theorem 89). Agievich [9] studied the correspondence between bent rectangles and regular q-ary bent functions [218], and described affine transformations of bent rectangles and presented the main known constructions of bent functions in terms of bent rectangles. Further development of this approach is likely to be useful.

A view on a hill

CHAPTER 7

Bent Functions with a Small Number of Variables

INTRODUCTION

In this chapter, we discuss what is known about bent functions in a small number of variables. Bent functions with not more than 14 variables are considered. We present extended affine classifications of bent functions in n variables and the exact numbers of them (up to $n = 8$), and give details of some other approaches to classification: in terms of trace forms, by bent rectangles, and by graphs of algebraic normal forms (for quadratic bent functions). Special bent functions (such as nonnormal functions) in a small number of variables are also considered. An overview of algorithms for bent functions generation is presented.

7.1 TWO AND FOUR VARIABLES

Two variables. The function $x_1 x_2$ is a representative of the unique class of extended affine equivalence. The set $\mathcal{B}_2$ consists of eight functions: Boolean functions with vectors of values having an odd Hamming weight.

Four variables. The set $\mathcal{B}_4$ consists of 896 Boolean functions, and all of them are quadratic. All bent functions in four variables are equivalent to the function $x_1 x_2 \oplus x_3 x_4$. The set $\mathcal{B}_4$ can be divided into 28 classes with 32 functions in every class. Algebraic normal forms (ANFs) of functions from a fixed class have the same quadratic part and differ from one another only by a linear part and a free term. Consider a graph on the set of variables such that an edge connects two variables if their product belongs to the ANF of the function. Then these 28 types can be given by graphs:

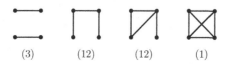

$$(3) \qquad (12) \qquad (12) \qquad (1)$$

Bent Functions
http://dx.doi.org/10.1016/B978-0-12-802318-1.00007-8

Near the graph one can see the number of types determined by it. For instance, there are three types of quadratic part with two items: $x_1x_2 \oplus x_3x_4$, $x_1x_3 \oplus x_2x_4$, and $x_1x_4 \oplus x_2x_3$. Two bent functions are called *graph equivalent* if their graphs are isomorphic.

7.2 SIX VARIABLES

Extended affine classification of bent functions in six variables was obtained by Rothaus [318]: the set $\mathcal{B}_6$ consists of four classes of extended affine equivalence. Representatives of all classes can be found in the following table.

N	Nonequivalent bent functions
1	$12 + 34 + 56$
2	$123 + 14 + 25 + 36$
3	$123 + 245 + 12 + 14 + 26 + 35 + 45$
4	$123 + 245 + 346 + 14 + 26 + 34 + 35 + 36 + 45 + 46$

For brevity, the function $x_1x_2 \oplus x_3x_4 \oplus x_5x_6$ is written as $12 + 34 + 56$.

Yang et al. [385] obtained the detailed algebraic classification. Let $\mathbb{F}_{2^6} = \{0, 1, \alpha, \alpha^2, \ldots, \alpha^{62}\}$, where α is a root of the polynomial $x^6 + x + 1$. Let a Boolean function f be identified by a function $f(c) : \mathbb{F}_{2^6} \to \mathbb{F}_2$, where c is an element of the field $\mathbb{F}_{2^6}$. Then as the representatives of the classes of extended affine equivalence for $\mathcal{B}_6$ one can choose the following functions: $\mathrm{tr}(c^3 + \alpha^5 c^5)$, $\mathrm{tr}(\alpha^3 c^7 + c^9)$, $\mathrm{tr}(\alpha c^3 + \alpha^6 c^7 + \alpha^{60} c^{13})$, and $\mathrm{tr}(c^7 + \alpha c^9 + c^{21})$, where tr is the trace function from $\mathbb{F}_{2^6}$ to $\mathbb{F}_2$.

Dillon [107] (see also [24]) proved that any bent function in six variables is equivalent to the function from the Maiorana-McFarland class (see Theorem 34). As mentioned in [33] all bent functions in six variables are normal (see Section 16.4).

The class $\mathcal{B}_6$ contains $5\,425\,430\,528 \simeq 2^{32.3}$ functions. This fact was obtained by Preneel [310]. A description of $\mathcal{B}_6$ in terms of bent squares—that is, bent rectangles when $r = k$ (see Theorem 31)—was given in 2000 by Agievich [7]. Let us say that two bent functions are *square equivalent* if the bent square of one of them can be obtained from the bent square of the other one by changing the signs of elements and by permutation of lines (rows, columns). Let $r = k = 3$. All functions from $\mathcal{B}_6$ break up into eight classes of square equivalence. Below one can see their representatives (bent squares of size $2^3 \times 2^3$) and cardinalities:

```
8  0  0  0  0  0  0  0
0  8  0  0  0  0  0  0
0  0  8  0  0  0  0  0
0  0  0  8  0  0  0  0
0  0  0  0  8  0  0  0
0  0  0  0  0  8  0  0
0  0  0  0  0  0  8  0
0  0  0  0  0  0  0  8
```

$$(2^{15} \cdot 3^2 \cdot 5 \cdot 7)$$

```
-4  4  4  4  0  0  0  0
 4 -4  4  4  0  0  0  0
 4  4 -4  4  0  0  0  0
 4  4  4 -4  0  0  0  0
 0  0  0  0  8  0  0  0
 0  0  0  0  0  8  0  0
 0  0  0  0  0  0  8  0
 0  0  0  0  0  0  0  8
```

$$(2^{18} \cdot 3 \cdot 7^2)$$

```
-4  4  4  4  0  0  0  0
 4 -4  4  4  0  0  0  0
 0  0 -4  4  4  4  0  0
 0  0  4 -4  4  4  0  0
 4  4  0  0 -4  4  0  0
 4  4  0  0  4 -4  0  0
 0  0  0  0  0  0  8  0
 0  0  0  0  0  0  0  8
```

$$(2^{21} \cdot 3 \cdot 7^2)$$

```
-4  4  4  4  0  0  0  0
 4 -4  0  0  4  4  0  0
 4  0 -4  0  4  0  4  0
 4  0  0 -4  0  4  4  0
 0  4  4  0  0  4 -4  0
 0  4  0  4 -4  0  4  0
 0  0  4  4  4 -4  0  0
 0  0  0  0  0  0  0  8
```

$$(2^{25} \cdot 3 \cdot 7)$$

```
-4  4  4  4  0  0  0  0
 4 -4  4  4  0  0  0  0
 4  4 -4  4  0  0  0  0
 4  4  4 -4  0  0  0  0
 0  0  0  0 -4  4  4  4
 0  0  0  0  4 -4  4  4
 0  0  0  0  4  4 -4  4
 0  0  0  0  4  4  4 -4
```

$$(2^{19} \cdot 7^2)$$

```
-4  4  4  4  0  0  0  0
 4 -4  4  4  0  0  0  0
 0  0 -4  4  4  4  0  0
 0  0  4 -4  4  4  0  0
 0  0  0  0 -4  4  4  4
 0  0  0  0  4 -4  4  4
 4  4  0  0  0  0 -4  4
 4  4  0  0  0  0  4 -4
```

$$(2^{20} \cdot 3^2 \cdot 7^2)$$

```
-4  4  4  4  0  0  0  0
 4 -4  0  0  4  4  0  0
 4  0 -4  0  4  0  4  0
 4  0  0 -4  0  4  4  0
 0  4  4  0 -4  0  0  4
 0  4  0  4  0 -4  0  4
 0  0  4  4  0  0 -4  4
 0  0  0  0  4  4  4 -4
```

$$(2^{23} \cdot 3 \cdot 7^2)$$

```
-6  2  2  2  2  2  2  2
 2 -6  2  2  2  2  2  2
 2  2 -6  2  2  2  2  2
 2  2  2 -6  2  2  2  2
 2  2  2  2 -6  2  2  2
 2  2  2  2  2 -6  2  2
 2  2  2  2  2  2 -6  2
 2  2  2  2  2  2  2 -6
```

$$(2^{23} \cdot 3^2 \cdot 5 \cdot 7)$$

In 2004, Meng et al. [264] enumerated $\mathcal{B}_6$ in another way (distinct from [7]). In 2013, the graph classification of quadratic bent functions in six variables was obtained by Korsakova [214]: there are exactly 45 graph nonequivalent bent functions, and 37 distinct types (the type is determined by the list of vertex degrees). They are as follows:

N	Type	Graph	N	Type	Graph
1	1 1 1 1 1 1		2	2 2 1 1 1 1	
3	2 2 2 2 1 1		4	3 2 2 1 1 1	
5	3 2 2 2 2 1		6	3 3 2 2 1 1	
7	3 3 2 2 2 2		8	3 3 3 1 1 1	
9	3 3 3 2 2 1		10	3 3 3 3 1 1	
11	3 3 3 3 2 2		12	4 2 2 2 1 1	
13	4 3 2 2 2 1		14	4 3 3 2 1 1	
15	4 3 3 2 2 2		16	4 3 3 3 2 1	
17	4 3 3 3 3 2		18	4 4 3 2 2 1	
19	4 4 3 3 1 1		20	4 4 3 3 2 2	

N	Type	Graph
21	4 4 3 3 3 1	
23	4 4 4 3 3 2	
25	4 4 4 4 3 3	
27	5 3 3 2 2 1	
29	5 3 3 3 3 3	
31	5 4 4 3 2 2	
33	5 4 4 4 4 1	
35	5 5 4 4 3 3	
37	5 5 5 5 5 5	

N	Type	Graph
22	4 4 3 3 3 3	
24	4 4 4 4 3 1	
26	5 2 2 2 2 1	
28	5 3 3 3 2 2	
30	5 4 3 3 2 1	
32	5 4 4 4 3 2	
34	5 5 3 3 3 3	
36	5 5 5 4 4 3	

7.3 EIGHT VARIABLES

Extended affine classification of bent functions in eight variables of degree not more than 3 was given by Hou [167] in 1998 (see also the PhD thesis of Braeken [24]). A paper by Agievich [8] is devoted to cubic bent functions of the special type. Bent functions in eight variables of degree not more than 3 break up into 10 classes of extended affine equivalence with the representatives given in the table below:

N	Nonequivalent bent functions of degree $\leqslant 3$
1	$12 + 34 + 56 + 78$
2	$123 + 14 + 25 + 36 + 78$
3	$123 + 245 + 34 + 26 + 17 + 58$
4	$123 + 245 + 13 + 15 + 26 + 34 + 78$
5	$123 + 245 + 346 + 35 + 26 + 25 + 17 + 48$
6	$123 + 245 + 346 + 35 + 13 + 14 + 27 + 68$
7	$123 + 245 + 346 + 35 + 26 + 25 + 12 + 13 + 14 + 78$
8	$123 + 245 + 346 + 35 + 16 + 27 + 48$
9	$127 + 347 + 567 + 14 + 36 + 25 + 45 + 78$
10	$123 + 245 + 346 + 147 + 35 + 27 + 15 + 16 + 48$

In 2006, Braeken [24] showed that all these functions are also equivalent to Maiorana-McFarland bent functions.

Lower ($2^{70.4}$) and upper ($2^{129.2}$) bounds for the size of the class $\mathcal{B}_8$ were obtained by Agievich [7] and Langevin et al. [229], respectively. Several results on the partial description of the class $\mathcal{B}_8$ based on the investigation of automorphism groups of bent functions can be found in papers by Dempwolff [103, 104]. In 2005, Yang et al. [385] showed that the set $\mathcal{B}_8$ consists of not fewer than 129 classes of extended affine equivalence. They give a list of representatives for these classes. These representatives are 53 functions of the form $\text{tr}(\alpha^i c^{d_1} + \alpha^j c^{d_2} + \alpha^k c^{d_3})$ and 76 functions of the form $\text{tr}(\alpha^i c^{d_1} + \alpha^j c^{d_2} + \alpha^k c^{d_3} + \alpha^\ell c^{d_4})$, where $\text{tr} : \mathbb{F}_{2^8} \to \mathbb{F}_2$ is the trace. In 2008, Gangopadhyay et al. [132] showed that the set $\mathcal{B}_8 \cap \mathcal{PS}$ contains functions from not fewer than six classes of extended affine equivalence, where $\mathcal{PS}$ is the partial spreads class (see Theorem 39).

Extended affine classification of bent functions in eight variables of degree 4 was achieved in 2008 and 2011 [225, 228]. Langevin and Leander described all 536 possible variants for the fourth-degree part of the ANF of a bent function in eight variables (here by the "ith-degree part" we mean the set of all items of the ANF that have degree i). The exact number of all bent functions in eight variables is given in [225, 228]: it is $2^9 \times 193\,887\,869\,660\,028\,067\,003\,488\,010\,240 \simeq 2^{106.29}$.

There are exactly $15\,104$ rotation-symmetric bent functions in eight variables, eight of them are also homogeneous—for details, see Section 16.3 or papers by Stănică and Maitra [339, 340].

7.4 TEN AND MORE VARIABLES

If $n \geqslant 10$, the class $\mathcal{B}_n$ is not described; the size of it is still unknown. In [385], the large set of bent functions in 10 variables is constructed; it is established that there are at least several hundred nonequivalent functions between them. One can find some information about the classes $\mathcal{B}_{10}$ and $\mathcal{B}_{12}$ in [104].

In 2006, Canteaut et al. [33] found some special bent functions with a small number of variables. They proved that there exist $\alpha, \beta \in \mathbb{F}_{2^{10}}$ such that a bent function $f(c) = \mathrm{tr}(\alpha c^{57} + \beta c)$ is nonnormal in 10 variables. For $n = 14$ they proved that a Kasami bent function $f(c) = \mathrm{tr}(\alpha c^{57})$ is nonweakly normal.

There are 12 homogeneous rotation-symmetric bent functions in 10 variables of degree 2; a list of them can be found in Section 16.3 or in a paper by Stănică and Maitra [340].

7.5 ALGORITHMS FOR GENERATION OF BENT FUNCTIONS

In this section, we cite several papers related to algorithms for generation of bent functions. Such algorithms are useful for constructing examples of bent functions in a relatively small number of variables.

Usually, the method for generating bent functions is based on one of the possible representations of a Boolean function and on special properties of this representation in the case of a bent function.

The algorithms proposed by Meng et al. [264, 265] are based on truth tables of bent functions (2004-2008). ANFs of bent functions were used in algorithms created by Fuller et al. [124, 125]. Spectral vectors of bent functions allowed Clark et al. [90] to create their algorithm in 2004. The algorithm of Yang et al. [385] was based on the trace representation of a bent function.

In the PhD thesis of Fuller [124], heuristic methods for constructing bent functions are considered in detail. The main idea of these methods is step-by-step changing of an initial Boolean function in order to improve its cryptographic properties (in particular, nonlinearity of a function). With this idea the following approaches were applied for generating bent functions: genetic algorithm [277], algorithms of a random search [124, 278], and simulated annealing [89]. Fuller [124] proposed a sufficiently fast algorithm for constructing pseudorandom bent functions of bounded degree. Starting with a quadratic bent function, the algorithm adds items of higher degrees to its ANF. To sort out a lot of "bad" items, combinatorial properties of bent functions are widely used.

In 2004-2008, Meng et al. [264] proposed an algorithm that can theoretically construct all bent functions in any number of variables. It is based on the relationship between the Walsh-Hadamard spectra of a Boolean function at partial points and the Walsh-Hadamard spectra of its subfunctions, and on the binary Möbius transform. Practically all bent functions in six variables were enumerated. With the restriction on the ANF, the algorithm is also efficient in the case of more variables. For example, enumeration of all homogeneous bent functions of degree 3 in 8 variables can be done in 1 min with a P4 1.7 GHz computer; the nonexistence of homogeneous bent functions in 10 variables of degree 4 is computationally proven in [264].

In 2000, Agievich [7] proposed an algorithm for generating distinct bent functions using bent squares (see Section 6.7). Let BS_n (*bent squares*) be the set of all bent functions in n variables generated by that algorithm. Then BS_n contains all Maiorana-McFarland functions (as bent functions with the simplest bent squares; see Theorem 35) and some other bent functions with bent squares that are more complicated. For each n the size of BS_n can be computed directly. This estimation gives the direct lower bound on the number of bent functions (see Theorem 89); for now it is the best known one.

Several aspects of generating random bent functions are discussed in a paper by A. Grocholewska-Czuryło [145]; see also [146]. For the generation of bent functions with random Boolean formulas, see Savicky [322]. Generation of bent sequences was discussed by Adams and Tavares [4].

7.6 CONCLUDING REMARKS

Classification of homogeneous rotation-symmetric bent functions in a small number of variables can be found in Section 16.3 or in [340].

Classification of bent functions in a small number of variables with respect to the local metrical equivalence is given in Section 11.5. There are exactly one, four, and six classes of locally metrically nonequivalent bent functions in four, six, and eight variables (up to degree 3), respectively.

Bounds for the number of bent functions in small number variables, $2 \leqslant n \leqslant 16$, can be found in Chapter 13.

CHAPTER 8

Combinatorial Constructions of Bent Functions

INTRODUCTION

It is very difficult not only to classify bent functions, but also to find methods for constructing them. Combinatorial methods for constructing bent functions such as iterative constructions, the Maiorana-McFarland construction, and the partial spread method are discussed in this chapter. Following work by Carlet, we divide constructions into two groups: *direct* and *iterative* (or *recursive*). The second group consists of methods that use already known bent functions (e.g., with fewer variables), whereas the first group contains only direct methods. We try to consider all known basic constructions of bent functions. We deal in general with *combinatorial* constructions, whereas in the next chapter *algebraic* constructions are considered.

8.1 ROTHAUS'S ITERATIVE CONSTRUCTION

Let n be an even integer in what follows. The first iterative construction was given by Rothaus [317, 318] in 1966.

Theorem 32 (Iterative construction). *A Boolean function $f(x, y) = g(x) \oplus h(y)$, where vectors x and y are of even lengths r and k, respectively, is a bent function in $r + k$ variables if and only if functions g and h are bent.*

This construction can be easily described in terms of bent rectangles [9]. Consider the following generalization introduced by Carlet [38]:

Theorem 33. *Let $n = r + k$, where r and k are even, and let f be a Boolean function in n variables. Let x and y run through $\mathbb{F}_2^r$ and $\mathbb{F}_2^k$, respectively. Suppose that functions $f_y(x) = f(x, y)$ are bent functions for all y. Define $g_x(y) = \tilde{f}_y(x)$. Then f is bent if and only if g_x is bent for every x.*

Note that Theorem 32 follows from Theorem 33.

Bent Functions
http://dx.doi.org/10.1016/B978-0-12-802318-1.00008-X

8.2 MAIORANA-MCFARLAND CLASS

This is one of the most important direct constructions of bent functions. It was introduced in 1973 [108, 260]. As mentioned by Dillon [107], J.A. Maiorana and R.L. McFarland came to this construction independently. An analog of this construction was proposed by V.A. Eliseev in 1962.

Theorem 34 (Maiorana-McFarland functions). *Let π be an arbitrary permutation on the set $\mathbb{F}_2^{n/2}$, and let h be an arbitrary Boolean function in $n/2$ variables. Then the function $f(x, y) = \langle x, \pi(y) \rangle \oplus h(y)$ is bent in n variables.*

Note that Rothaus [317, 318] proved this theorem in the case of identical π. From Theorem 34 it is easy to conclude that there are bent functions of any degree d such that $2 \leqslant d \leqslant n/2$.

Sometimes the *completed Maiorana-McFarland class* of bent functions is considered. The means that together with Maiorana-McFarland bent functions we deal with all bent functions extended affine equivalent to them (see Section 5.2).

One can prove this theorem as an exercise. The main idea of this construction is the "concatenation of affine functions" as mentioned by Carlet [46]. Indeed, for every fixed value for the second part of the variables, function f is affine in the first $n/2$ variables. On the other hand, affine functions arise when we consider the associated bent squares (see the definition in Section 6.7). In 2000, Agievich proved [7] the following theorem:

Theorem 35. *A bent function belongs to the Maiorana-McFarland class if and only if all lines (rows and columns) of its bent square are spectral vectors of affine Boolean functions.*

In 2006, Canteaut et al. [33] proved a very nice property of Maiorana-McFarland bent functions (as we will see in Section 16.4, these functions are "very normal"); namely, the following theorem:

Theorem 36. *A Boolean function f in n variables is extended affine equivalent to a Maiorana-McFarland bent function if and only if there is an affine subspace of dimension $n/2$ such that f is affine on each of its cosets.*

In Theorem 34, the variables of function f are divided into halves. In 2004, Carlet [41] (see also [46]) generalized the idea of McFarland and considered the partition of variables into nonequal parts.

Theorem 37. *Let $n = r + k$ and let $h : \mathbb{F}_2^k \to \mathbb{F}_2^r$ be an arbitrary mapping such that for any vector x of length r the set $h^{-1}(x)$ is a subspace of dimension $n - 2r$ in $\mathbb{F}_2^k$. Let g be a Boolean function in k variables such that the section of it on the set $h^{-1}(x)$ for any x is a bent function for $n > 2r$. Then a Boolean function $f(x, y) = \langle x, h(y) \rangle \oplus g(y)$ is bent in n variables.*

Note that the construction of Carlet is very similar to the method of describing bent functions proposed by Yashchenko [386] in 1997 (remember Theorem 28). Theorem 34 is a particular case of Theorem 37 if $r = k = n/2$.

A q-ary analog of the Maiorana-McFarland construction can be found in Section 15.2. Note that the $\mathbb{Z}_4$ version of Maiorana-McFarland functions was considered by Wolfmann [378] in 1988.

In 2007, Weng et al. [377] proposed some upper and lower bounds on the ranks of Maiorana-McFarland bent functions. It was proven that almost every Desarguesian partial spread bent function is not equivalent to any Maiorana-McFarland bent function. Later, in 2008, Weng et al. [376] introduced new upper and lower bounds on the ranks of Maiorana-McFarland bent functions; bent functions that achieve these bounds were determined.

Note that the dual function to the Maiorana-McFarland bent function $f(x, y) = \langle x, \pi(y) \rangle \oplus h(y)$ has the form $f(x, y) = \langle \pi^{-1}(x), y \rangle \oplus h(\pi^{-1}(x))$; see [46]. Self-dual Maiorana-McFarland bent functions were characterized by Carlet et al. [52] (2008, 2010). They proved the following theorem:

Theorem 38. *A bent function $f(x, y) = \langle x, \pi(y) \rangle \oplus h(y)$ is self-dual (or anti-self-dual) if and only if $h(y) = \langle b, y \rangle \oplus \epsilon$ and $\pi(y) = L(y) \oplus a$, where L is a linear automorphism satisfying $L \times L^t = I_{n/2}$, $a = L(b)$, and a has an even (or odd) Hamming weight.*

The relation to codes is also presented in that paper.

In 2011, Kolokotronis and Limniotis [204] considered Maiorana-McFarland functions with high second-order nonlinearity.

In 2013, Gangopadhyay [127] studied nonequivalence of cubic Maiorana-McFarland bent functions.

Maiorana-McFarland construction of generalized bent functions was proposed by Stănică et al. [342]; see also Section 15.6.

8.3 PARTIAL SPREADS: $\mathcal{PS}^+, \mathcal{PS}^-$

The next direct construction of Dillon [108], obtained in 1974, is based on the special sets of subspaces in an n-ary Boolean cube. It has the name *partial spreads*. Although the construction is direct, it is difficult to give a concrete bent function obtained with it.

Let $\text{Ind}_S : \mathbb{F}_2^n \to \mathbb{F}_2$ be the characteristic function of a subset $S \subseteq \mathbb{F}_2^n$.

Theorem 39. *Let q be equal to $2^{(n/2)-1}$ or $2^{(n/2)-1} + 1$. Let $L_1, \ldots, L_q$ be linear subspaces of dimension $n/2$ of the space $\mathbb{F}_2^n$ such that any two of them have in an intersection only zero vector. Then the function $f(x) = \bigoplus\limits_{i=1}^{q} \mathrm{Ind}_{L_i}(x)$ is a bent function in n variables.*

All bent functions that can be constructed by Theorem 39 with $q = 2^{(n/2)-1}$ form the class $\mathcal{PS}^-$ of bent functions. The class $\mathcal{PS}^+$ is defined by the value $q = 2^{(n/2)-1} + 1$.

Together classes $\mathcal{PS}^-$ and $\mathcal{PS}^+$ form the class $\mathcal{PS}$. It is known [46] that all bent functions from $\mathcal{PS}^-$ have the maximal possible degree $n/2$. The degrees of bent functions from $\mathcal{PS}^+$ can be different. For instance, there are quadratic bent functions in $\mathcal{PS}^+$.

There is a problem how to find a constructive method for obtaining bent functions from $\mathcal{PS}$, to describe algebraic normal forms of such bent functions. Dillon proposed a way how to obtain examples from $\mathcal{PS}^-$. In [108], he constructed the subclass $\mathcal{PS}_{\mathrm{ap}} \subset \mathcal{PS}^-$; it will be considered in the next section.

Generalized partial spreads were discussed by Carlet [37] in 1995, and later by Guillot [147] in 2001.

In 2007, Weng et al. [377] studied the ranks of partial spread bent functions. It was proven that almost every Desarguesian partial spread bent function is not equivalent to any Maiorana-McFarland bent function.

In 2014, Lisonek and Lu [239] used partial spreads of $\mathbb{F}_p^n$ to construct two classes of bent functions from $\mathbb{F}_p^n$ to $\mathbb{F}_p$. Their constructions generalize classes $\mathcal{PS}^-$ and $\mathcal{PS}^+$.

8.4 DILLON'S BENT FUNCTIONS: $\mathcal{PS}_{\mathrm{ap}}$

As already mentioned, Dillon [108] proposed a simple way to obtain examples from $\mathcal{PS}^-$.

Let $\mathbb{F}_2^n$ be viewed as a two-dimensional vector space over $\mathbb{F}_{2^{n/2}}$—that is, consider $\mathbb{F}_2^n \approx \mathbb{F}_{2^{n/2}} \times \mathbb{F}_{2^{n/2}}$. Then a Boolean function in n variables can be represented as $f : (x, y) \to z$, where $x, y \in \mathbb{F}_{2^{n/2}}$, $z \in \mathbb{F}_2$. Note that this two-dimensional space can be represented as the union of $2^{n/2} + 1$ lines through 0 such that any two of them intersect only by a zero element. These lines are $n/2$-dimensional subspaces of $\mathbb{F}_2^n$ over $\mathbb{F}_2$. Then choose any $2^{n/2-1}$ of these lines and take them to be different from those of equations $x = 0$ and $y = 0$. Then we get an example of $\mathcal{PS}^-$ belonging to $\mathcal{PS}_{\mathrm{ap}}$. The following theorem holds:

Theorem 40. *Let g be an arbitrary balanced Boolean function in $n/2$ variables such that $g(0) = 0$. Then the Boolean function f in n variables defined by $f(x, y) = g(xy^{2^{n/2}-2})$, for all $x, y \in \mathbb{F}_{2^{n/2}}$, is a bent function.*

Note that $f(x, y) = g(xy^{2^{n/2}-2})$ can also be written as $f(x, y) = g(\frac{x}{y})$ if we assume $\frac{x}{y} = 0$ when $y = 0$. Bent functions constructed by Theorem 40 form the subclass $\mathcal{PS}_{ap}$ of the class $\mathcal{PS}^-$. In the literature, bent functions constructed via Theorem 40 are usually called *Dillon bent functions*.

In 2008-2010, Carlet et al. [52] proved the following theorem:

Theorem 41. *A Dillon function g is self-dual bent if g satisfies $g(1) = 0$ and $g(y) = g(1/y)$ for all nonzero y. There are exactly $\binom{2^{n/2-1} - 1}{2^{n/2-2}}$ such functions.*

In 2013, Gangopadhyay et al. [129] studied the upper bounds on algebraic immunity of bent functions from $\mathcal{PS}_{ap}$.

8.5 DOBBERTIN'S CONSTRUCTION

In 1995, Dobbertin [111] proposed a direct construction of bent functions that unifies Maiorana-McFarland and Dillon bent functions as two opposite extremal cases.

Theorem 42. *Let g be a balanced Boolean function on $\mathbb{F}_{2^{n/2}}$, and let $\phi, \psi : \mathbb{F}_{2^{n/2}} \rightarrow \mathbb{F}_{2^{n/2}}$ be such that ϕ is one-to-one and ψ is arbitrary. Then the function*

$$f_{g,\phi,\psi}(x, \phi(y)) = \begin{cases} g(\frac{x+\psi(y)}{y}), & \text{if } y \neq 0, \\ 0, & \text{otherwise} \end{cases}$$

is bent in n variables if and only if (g, ϕ, ψ) is a bent triple.

There are several possibilities for (g, ϕ, ψ) to be a bent triple. For example, if g is affine and ϕ, ψ are arbitrary, we have a bent triple. Bent functions constructed in this case coincide with Maiorana-McFarland functions. For other variants when (g, ϕ, ψ) is a bent triple, see [111].

8.6 MORE ITERATIVE CONSTRUCTIONS

The next iterative construction was obtained by Rothaus [317] and Dillon [108].

Theorem 43. *Let f', f'', and f''' be bent functions in n variables such that their sum is bent again. Then $g(a_1, a_2, x) = f'(x)f''(x) \oplus f'(x)f'''(x) \oplus f''(x)f'''(x) \oplus a_1 f'(x) \oplus a_1 f''(x) \oplus a_2 f'(x) \oplus a_2 f'''(x) \oplus a_1 a_2$ is a bent function in $n + 2$ variables.*

But, in general, it is not clear when for distinct collections $\{f_1', f_1'', f_1'''\}$ and $\{f_2', f_2'', f_2'''\}$ one can obtain distinct bent functions g.

Another construction was introduced by Carlet [43].

Theorem 44. *Suppose that f', f'', and f''' are bent functions in n variables such that their sum (denote it by s) is bent. Moreover, let $\tilde{s} = \tilde{f}' \oplus \tilde{f}'' \oplus \tilde{f}'''$. Then function $g(x) = f'(x)f''(x) \oplus f'(x)f'''(x) \oplus f''(x)f'''(x)$ is bent in n variables.*

Note that checking the conditions (when s is bent and when $\tilde{s} = \tilde{f}' \oplus \tilde{f}'' \oplus \tilde{f}'''$) is difficult enough.

In 2013, the following constructions were obtained by Korsakova [214] during her study of distinct methods to obtain bent functions in six variables from bent functions in four variables:

Theorem 45. *Let f be a bent function in n variables. Then*

(1) for any $\alpha, \beta \in \mathbb{F}_2$ the function $g(x, x_{n+1}, x_{n+2}) = f(x) \oplus \alpha \cdot x_i x_{n+1} \oplus \beta \cdot x_i x_{n+2} \oplus x_{n+1} x_{n+2}$ is a bent function in $n + 2$ variables;

(2) for any $i, j \in \{1, \ldots, n\}$ the function $g(x, x_{n+1}, x_{n+2}) = f(x) \oplus x_i x_{n+1} \oplus x_j x_{n+1} \oplus x_{n+1} x_{n+2}$ is bent in $n + 2$ variables.

In 2012, Carlet et al. [69] also studied iterative constructions of bent functions and their enforcement.

8.7 MINTERM ITERATIVE CONSTRUCTIONS

Climent et al. [91, 93] suggested in 2008 and 2014 the following iterative constructions of bent functions in $n + 2$ variables from bent functions in n variables. A key role for these constructions is played by minterms.

A *minterm* in n variables $x_1, \ldots, x_n$ is a Boolean function defined as

$$m_{(u_1, u_2, \ldots, u_n)}(x_1, x_2, \ldots, x_n) = (1 \oplus u_1 \oplus x_1)(1 \oplus u_2 \oplus x_2) \ldots (1 \oplus u_n \oplus x_n),$$

where $(u_1, u_2, \ldots, u_n)$ is a binary vector. Denote by $m_i(x)$ a *simple minterm*, $i = 0, 1, \ldots, 2^n - 1$; namely, such a function that takes 1 only on the vector that is a binary representation of i.

The following theorem was proven in [91]:

Theorem 46. *The following statements hold:*

(1) Let f be a bent function in n variables. Let m_i be a simple minterm in two variables. Then $g(x, x_{n+1}, x_{n+2}) = f(x) \oplus m_i(x_{n+1}, x_{n+2})$ is a bent function in $n + 2$ variables.

(2) Let f' and f'' be two distinct bent functions in n variables, such that $f' \oplus f'' \neq 1$. Let σ be an arbitrary permutation on $\{0, 1, 2, 3\}$. Then a Boolean function $g(x, x_{n+1}, x_{n+2}) = f'(x)(m_{\sigma(0)}(x_{n+1}, x_{n+2}) \oplus m_{\sigma(1)}(x_{n+1}, x_{n+2})) \oplus$

$f''(x)m_{\sigma(2)}(x_{n+1}, x_{n+2}) \oplus (f''(x) \oplus 1)m_{\sigma(3)}(x_{n+1}, x_{n+2})$ *is   a   bent function in $n + 2$ variables.*

Using this theorem, one can construct $6|\mathcal{B}_n|^2 - 8|\mathcal{B}_n|$ distinct bent functions in $n + 2$ variables. We recall this result in Chapter 13 (Theorem 90).

In 2014, Climent at al. [93] proved the following theorem:

Theorem 47. *Let f be a bent function in n variables, and let u and v be arbitrary vectors of length n. Let σ be a permutation on $\{0, 1, 2, 3\}$. Then the following Boolean function in $n + 2$ variables is bent:*

$$g(x, x_{n+1}, x_{n+2}) = \bigoplus_{x_{n+1}, x_{n+2} \in \mathbb{F}_2} f(x \oplus x_{n+1} \cdot v \oplus x_{n+2} \cdot u) \cdot m_{\sigma(2x_{n+1}+x_{n+2})}.$$

Then $6|\mathcal{B}_n|^2 + 2^{n+2}(2^n - 3)|\mathcal{B}_n|$ distinct bent functions in $n + 2$ variables can be constructed using Theorems 46 and 47. When $n \geqslant 6$, this number is not better than the number following from Theorem 46.

8.8 BENT ITERATIVE FUNCTIONS: $\mathcal{BI}$

An extensive study of the restrictions of bent functions to affine subspaces was proposed by Canteaut and Charpin [31] in 2003. In particular, they have established that restrictions of a bent function f to a subspace V of codimension 2 and to its cosets are bent if and only if the derivative of $\tilde{f}$ with respect to $V^\perp$ is constant equal to 1. This last result can be interpreted as an iterative construction for bent functions.

A hoarfrost

In this section, we present this simplified variant of the iterative construction of Canteaut and Charpin [31] and equip it with a new proof. This proof was obtained in [363], in which bent iterative functions were used for evaluating the number of all bent functions.

Let a Boolean function g in $n + 2$ variables be defined as

$$g(00, x) = f_0(x), \quad g(01, x) = f_1(x), \tag{8.1}$$

$$g(10, x) = f_2(x), \quad g(11, x) = f_3(x),$$

where f_0, f_1, f_2, and f_3 are Boolean functions in n variables. Note that for distinct ordered collections $\{f_0, f_1, f_2, f_3\}$ we always obtain distinct functions g. From [31] the next theorem follows:

Theorem 48. *Let functions f_0, f_1, and f_2 be bent functions in n variables. Then function g defined by (8.1) is a bent function in $n + 2$ variables if and only if f_3 is a bent function in n variables and $\tilde{f_0} \oplus \tilde{f_1} \oplus \tilde{f_2} \oplus \tilde{f_3} = 1$.*

Proof. ($\Longleftarrow$) Let f_0, f_1, f_2, and f_3 be bent functions and let $\tilde{f_0} \oplus \tilde{f_1} \oplus \tilde{f_2} \oplus \tilde{f_3} = 1$. Show that g is bent. We have

$$W_g(a_1, a_2, x) = W_{f_0}(x) + (-1)^{a_2} W_{f_1}(x) + (-1)^{a_1} W_{f_2}(x)$$
$$+ (-1)^{a_1 \oplus a_2} W_{f_3}(x).$$

Using dual functions, we obtain

$$W_g(a_1, a_2, x) = 2^{n/2} \left((-1)^{\tilde{f_0}(x)} \oplus (-1)^{a_2 \oplus \tilde{f_1}(x)} \right.$$
$$\left. + (-1)^{a_1 \oplus \tilde{f_2}(x)} \oplus (-1)^{a_1 \oplus a_2 \oplus 1 \oplus \tilde{f_0}(x) \oplus \tilde{f_1}(x) \oplus \tilde{f_2}(x)} \right).$$

The possible values for the expression between parentheses are ± 4, ± 2, and 0. In fact this expression is always equal to ± 2. Indeed, in the case $\tilde{f_0}(x) = \tilde{f_1}(x) = \tilde{f_2}(x) = 0$, we obtain the expression

$$R(a_1, a_2) = 1 + (-1)^{a_2} + (-1)^{a_1} + (-1)^{a_1 \oplus a_2 \oplus 1},$$

which for any a_1, a_2 equals ± 2. It is easy to prove that in all other (seven) cases for $\tilde{f_0}(x), \tilde{f_1}(x), \tilde{f_2}(x)$ the expression in parentheses differs from $R(a_1, a_2)$ only by changing the signs for even number of items. Hence it can be equal to ± 2 or $\pm 2 \pm 4$ only. Since it is not more than 4, only one possible value, ± 2, remains. Thus, for any a_1, a_2, x, we have $W_g(a_1, a_2, x) = \pm 2^{(n+2)/2}$, and therefore g is a bent function.

($\Longrightarrow$) Let g be a bent function. Then its Walsh-Hadamard coefficient

$$W_g(a_1, a_2, x) = 2^{n/2} \left((-1)^{\tilde{f_0}(x)} + (-1)^{a_2 \oplus \tilde{f_1}(x)} \right.$$

$$\left. + (-1)^{a_1 \oplus \tilde{f_2}(x)} + (-1)^{a_1 \oplus a_2} \frac{W_{f_3}(x)}{2^{n/2}} \right)$$

is equal to $\pm 2^{(n+2)/2}$ for all a_1, a_2, x. It is obvious that the expression in parentheses should be equal to ± 2. The necessary condition for it is that the fourth item in parentheses has to be a natural number. But according to Parseval's equality for W_{f_3} it is true if and only if $W_{f_3}(x) = \pm 2^{n/2}$ for all x—that is, f_3 is a bent function. Thus, the fourth item has the form $(-1)^{x_1 \oplus x_2 \oplus \tilde{f_3}(x)}$.

It is easy to see now that the value $\tilde{f_3}(x)$ has to be defined by the values $\tilde{f_0}(x), \tilde{f_1}(x),$ and $\tilde{f_2}(x)$ in a unique way. It is true since the sum in parentheses is equal to ± 2. It remains to be noted that we have already found this appropriate way for $\tilde{f_3}(x)$ to be defined. Namely, $\tilde{f_3} = \tilde{f_0} \oplus \tilde{f_1} \oplus \tilde{f_2} \oplus 1$. $\square$

Let us give some examples. Here Boolean functions are presented by their vectors of values.

- *The bent iterative function* $g = (0001\ 0001\ 0001\ 1110)$ *is obtained by taking* $f_0 = f_1 = f_2 = (0001)$. *Note that* $\tilde{f_0} = f_0$. *The function* f_3 *can be found from the equality* $\tilde{f_3} = \tilde{f_0} \oplus \tilde{f_1} \oplus \tilde{f_2} \oplus 1 = (1110)$. *Note also that here* $\tilde{f_3} = f_3$.
- *The bent iterative function* $g = (0001\ 0010\ 0001\ 1101)$ *is constructed by taking* $f_0 = (0001)$, $f_1 = (0010)$, *and* $f_2 = (0001)$. *Note that again* $\tilde{f_0} = f_0 = f_2$, *but* $\tilde{f_1} = (0100)$. *The function* f_3 *is derived from* $\tilde{f_3} = \tilde{f_0} \oplus \tilde{f_1} \oplus \tilde{f_2} \oplus 1 = (1011)$. *Then* $f_3 = (1101)$.
- *The bent iterative function* $g = (0010\ 0001\ 0001\ 1101)$ *is constructed by taking* $f_0 = (0010)$, $f_1 = (0001)$, *and* $f_2 = (0001)$. *Here* $\tilde{f_0} = (0100)$, $\tilde{f_1} = f_1 = f_2$. *The function* f_3 *is obtained from* $\tilde{f_3} = \tilde{f_0} \oplus \tilde{f_1} \oplus \tilde{f_2} \oplus 1 = (1011)$. *Then* $f_3 = (1101)$.

Note that three distinct bent functions in n variables can produce up to six distinct bent functions in $n + 2$ variables, since it is possible to order them in 3! ways.

Bent functions that can be obtained by Theorem 48 are called *bent iterative functions*. Let $\mathcal{BI}_{n+2}$ (*bent iterative*) denote the class of all such functions in $n + 2$ variables. This class of bent functions is studied further in Chapter 13 with respect to the lower bound on the number of all bent functions.

Note that according to Canteaut and Charpin [31] there are bent functions from the Maiorana–McFarland class and from the class $\mathcal{PS}$ that cannot be represented as bent iterative functions. Also, as follows from the investigation of nonnormal functions by Canteaut et al. [33] there are bent functions in $\mathcal{BI}_n$ that are nonequivalent to Maiorana-McFarland bent functions.

8.9 OTHER CONSTRUCTIONS

The following fact can be found in an article by Hou and Langevin [175] from 1997:

Theorem 49. *Let f be a Boolean function in n variables, and let h be a permutation on the set $\mathbb{F}_2^n$. Denote by $h_1, \ldots, h_n$ such Boolean functions where $h(x) = (h_1(x), \ldots, h_n(x))$. A function $f \circ h^{-1}$ is bent if for any γ*

$$\mathrm{dist}\left(f, \bigoplus_{i=1}^{n} \gamma_i h_i\right) = 2^{n-1} \pm 2^{(n/2)-1}.$$

Some historical constructions of bent functions and difference sets can be found in a paper by Nyberg [287] published in 1991. A method for direct construction of bent functions can be found in the PhD thesis of Stănică [335] from 1998. In 1999, Xiang [384] considered a construction of bent functions in $2n$ variables (n is odd) that uses maximally nonlinear functions. Aspects of designing bent functions and resilient functions from known ones without extending the number of variables were discussed by Carlet [42]. In 2005, Carlet and Yucas [68] proposed some piecewise constructions of bent and almost optimal Boolean functions.

In 2007, a construction of a bent function $(\gamma \oplus 1)p(x) \oplus \gamma q(x)$ obtained from two Boolean functions $p(x)$ and $q(x)$ was proposed by Zhao [397]. He also studied the iterative lower bound on the number of bent functions. Later, in 2014, construction of the same form was discussed by Wolfmann [381]—unfortunately, without reference to the paper by Zhao [397].

Constructing bent functions by using generalized Hadamard matrices was discussed by Wang and Xu [375] in 2009. In 2012, Climent et al. [92] proposed a construction of bent functions in n variables from a basis of $\mathbb{F}_2^n$.

We also mention several papers related to constructing bent functions via some known objects of discrete mathematics. The question how to construct bent functions from plateaued functions was discussed in 2013 by Çeşmelioğlu and Meidl [75]. Earlier, in 2011, Carlet [48] considered how to construct almost perfect nonlinear functions from bent functions (vectorial).

CHAPTER 9

Algebraic Constructions of Bent Functions

INTRODUCTION

In this chapter, another type of construction of bent functions is considered. This type is called *algebraic* since bent functions are considered as functions from $\mathbb{F}_{2^n}$ to $\mathbb{F}_2$ and their constructions are given in trace form. First, we discuss several details of the algebraic approach to the construction of bent functions. Bent functions of the form $f(c) = \mathrm{tr}(ac^d)$ are called *monomial bent functions*. The number d is a *bent exponent*. Five known bent exponents proposed by Gold; Dillon; Kasami; Canteaut and Leander; and Canteaut, Charpin, and Kuyreghyan are studied. We collect results related to these monomial bent functions such as information about the parameters for which the constructions work, relations to other known classes of bent functions, and special properties of monomial bent functions (e.g., k-essential dependence of Kasami bent functions). Some details on Niho exponents for bent functions are given. Finally, we discuss the possibility of a general algebraic approach for classification of bent functions.

9.1 AN ALGEBRAIC APPROACH

The first series of algebraic constructions are called *power* or *monomial bent functions*. Let vectorial space $\mathbb{F}_2^n$ be identified with Galois field $\mathbb{F}_{2^n}$. Recall that Boolean functions in n variables can be considered as functions from $\mathbb{F}_{2^n}$ to $\mathbb{F}_2$ since an appropriate element of $\mathbb{F}_{2^n}$ can correspond to a fixed vector c (see Section 1.7). We denote this element of the field also denote by c. In this chapter, we always consider Boolean functions in n variables.

Recall that $\mathrm{tr} : \mathbb{F}_{2^n} \to \mathbb{F}_2$ is the *trace*—that is, $\mathrm{tr}(c) = c + c^2 + \cdots + c^{2^{n-1}}$.

Bent functions of the form

$$f(c) = \mathrm{tr}(ac^d),$$

Bent Functions
http://dx.doi.org/10.1016/B978-0-12-802318-1.00009-1

where $a \in \mathbb{F}_{2^n}^*$ is a certain parameter, are called *monomial*. The number d is a *bent exponent*. Monomial bent functions are foremost interesting for cryptographic applications. There are five known bent exponents: Gold, Dillon, Kasami, Canteaut-Leander (MF-1), and Canteaut-Charpin-Kuyreghyan (MF-2). We consider them in the next sections. Note that the following questions in this direction are still open:

- *Are there bent functions with other exponents?* It is known [226] that up to $n = 24$ the answer to this question is *no*.
- *Is it possible to find a simple combinatorial construction for a monomial bent function?*

Details of monomial constructions of bent functions are discussed in papers by Leander [230] (with some corrections provided by Sun and Wu [347]), Leander and Langevin [226], and others.

The second series of bent functions contains functions of the form

$$f(c) = \mathrm{tr}(a_1 c^{d_1} + a_2 c^{d_2})$$

for appropriate elements $a_1, a_2 \in \mathbb{F}_{2^n}$ and exponents d_1, d_2. There are known examples of such functions with special exponents—*Niho exponents* of the form

$$d \equiv 2^i \mod 2^{n/2} - 1.$$

We consider them in Section 9.8.

Some ideas on the general algebraic approach to the construction of bent functions are discussed in Section 9.9.

9.2 BENT EXPONENTS: GENERAL PROPERTIES

In this section, we gather some known necessary conditions for a number d to be a bent exponent. Here, we follow Leander [230].

Recall that Boolean functions in n variables are considered.

Let d be a bent exponent. By Theorem 16 (degree of a bent function) and Theorem 10 (degree of a monomial function), one can easily find that the binary weight of d should be at most $n/2$.

Since a bent function is not balanced (see Theorem 22), d cannot be coprime to $2^n - 1$. Otherwise mapping $c \to ac^d$ is one-to-one (if $a \neq 0$) and $\sum_{c \in \mathbb{F}_{2^n}} \mathrm{tr}(ac^d) = \sum_{c \in \mathbb{F}_{2^n}} \mathrm{tr}(c)$. As far as $\mathrm{tr}(c)$ is a linear nonconstant function and hence balanced, this sum is equal to 0. Then $\mathrm{tr}(ac^d)$ should be balanced, but this contradicts the property of being bent.

Canteaut [30] proved that d must be coprime either to $2^{n/2} - 1$ or to $2^{n/2} + 1$.

Let $\gcd(\cdot, \cdot)$ be the greatest common divisor of two numbers. In general, the following theorem holds:

Theorem 50. *Let d be a bent exponent of a monomial bent function in n variables. Then it holds that*

(1) $\text{wt}(d) \leqslant n/2$;

(2) $\gcd(d, 2^n - 1) > 1$;

(3) $\gcd(d, 2^{n/2} + 1) = 1$ *if and only if* $\tilde{f}(0) = 0$;

(4) $\gcd(d, 2^{n/2} - 1) = 1$ *if and only if* $\tilde{f}(0) = 1$.

Theorem 51. *Let d be a bent exponent of a monomial bent function in n variables. Then*

(1) *there exist a such that $\text{tr}(ac^d)$ is not bent;*

(2) *if $ab^{-1} \in \{c^d \mid c \in \mathbb{F}_{2^n}\}$, then functions $\text{tr}(ac^d)$ and $\text{tr}(bc^d)$ are linear equivalent.*

For more details, see [230] (with some corrections provided in [347]) and Carlet [46].

9.3 GOLD BENT FUNCTIONS

Consider the simplest monomial construction for bent functions. Recall that we are dealing with Boolean functions in n variables, where n is even.

Theorem 52. *The number $d = 2^i + 1$, where $\frac{n}{\gcd(n,i)}$ is even and i is a natural number, is a bent exponent.*

Gold bent functions are quadratic.

It is known that three types of monomial bent functions can be described in terms of the Maiorana-McFarland construction. These bent functions are the Gold, Canteaut-Leander (Section 9.6), and Canteaut-Charpin-Kuyreghyan (Section 9.7) bent functions.

More details follow from the next theorem proved in [230]:

Theorem 53. *The function $f(c) = \text{tr}(\alpha c^d)$, where $d = 2^i + 1$, where i is a natural number, is bent if and only if $\alpha \notin \{c^d \mid c \in \mathbb{F}_{2^n}\}$. The dual of function f is given by $\tilde{f}(c) = f(h^{-1}(c^{2^i})) + \tilde{f}(0)$, where $h(c) = \alpha^{2^i} c^{2^{2i}} + \alpha c$.*

From the first part of this theorem it is easy to conclude that for the Gold exponent there is a coefficient α such that f is bent if and only if $c \to c^d$ is not a one-to-one function.

In 2007, Lahtonen et al. [224] proved that the Gold functions yield bent functions when they are restricted to certain hyperplanes.

9.4 DILLON EXPONENT

This construction was obtained by Dillon [108] in 1974 as an example of bent functions belonging to the class $\mathcal{PS}$ (see Theorem 39).

Theorem 54. *The number $d = 2^{n/2} - 1$ is a bent exponent.*

Dillon [108] proved the following fact about the choice of the coefficient α:

Theorem 55. *Let α be an element of subfield $E = \mathbb{F}_{2^{n/2}}$ of $\mathbb{F}_{2^n}$. A Boolean function $f(c) = \text{tr}(\alpha c^d)$, where $d = 2^{n/2} - 1$, is bent if and only if α is a 0 of the Kloosterman sum*

$$K(\alpha) = \sum_{x \in E} (-1)^{\text{tr}_E(x^{-1} + \alpha x)}.$$

Here $\text{tr}_E(c) = c + c^2 + \cdots + c^{2^{(n/2)-1}}$ is a trace from the subfield E to $\mathbb{F}_2$.

In 2013, Li et al. [234] proposed several new classes of bent functions from Dillon exponents.

9.5 KASAMI BENT FUNCTIONS

Let us consider the most interesting construction of monomial bent functions.

Theorem 56. *The number $d = 2^{2k} - 2^k + 1$, where k is an integer and $\gcd(k, n) = 1$, is a bent exponent.*

A Boolean function in n variables (n is even) $f(c) = \text{tr}(\alpha c^d)$ is called *a Kasami Boolean function* if $d = 2^{2k} - 2^k + 1$, $\gcd(k, n) = 1$, $\alpha \in \mathbb{F}_{2^n}^*$.

From Theorem 57 it is clear how to choose a coefficient α for a Kasami Boolean function to be bent. Theorem 57 was conjectured by Hollmann and Xiang [166] in 1999 and was proven by Dillon and Dobbertin [109] in 2004 for the case when $\gcd(n, 3) = 1$. An alternative proof of the theorem was given by Langevin and Leander [226]; they proved also that the restriction $\gcd(n, 3) = 1$ is unnecessary.

Theorem 57. *Let n be even. A Kasami Boolean function is bent if and only if α does not belong to the set $\{c^3 | c \in \mathbb{F}_{2^n}\}$.*

Further by "Kasami bent functions" we mean exactly those mentioned in Theorem 57.

Kasami bent functions are the most complicated from the class of monomial bent functions. The algebraic degree of them can be equal to any even number from 2 to $n/2$. It is known that bent functions dual to Kasami bent functions are not monomial [226]. The class of Kasami bent functions contains nonnormal and nonweakly normal functions [33]. For example, all Kasami bent functions in six variables are normal, but for $n = 14$ it has been proven [33] that a Kasami bent function $f(c) = \text{tr}(\alpha c^{57})$ is nonweakly normal (and hence nonnormal too). In fact, any Kasami bent function in n variables is normal if 6 divides n [33].

Also there are Kasami bent functions that are not equivalent to functions from the Maiorana-McFarland and $\mathcal{PS}$ classes; see papers by Sharma and Gangopadhyay [325] and Canteaut et al. [33].

In 2007, Lahtonen et al. [224] studied Walsh-Hadamard spectra of Kasami functions using elementary facts about quadratic forms in characteristic 2.

In 2008, Langevin and Leander proved [226] that a bent function dual to a Kasami bent function is not monomial. The wrote: "We presently cannot answer the more interesting question whether the dual is linear or affine equivalent to a monomial bent function" (in particular, to a Kasami bent function). That is why a result declared in [227] still remains a hypothesis.

In 2013, combinatorial properties of Kasami bent functions were studied by Frolova [123]. She proved that multiple derivatives of high order for Kasami bent functions are not identically equal to 0. Thus, it follows that Kasami bent functions are essentially dependent on the products of their variables. Consider this result in more detail.

Recall that a Boolean function $D_a f(c) = f(c) + f(c + a)$ is called a *derivative of f with respect to a*, $a \in \mathbb{F}_{2^n}$. Sharma and Gangopadhyay [325] showed that for any a, b from $\mathbb{F}_{2^n}$ the second derivative $D_a D_b f(c)$ of a Kasami bent function $f(c)$ in n variables is not identically equal to 0. Frolova [123] proved the following theorem:

Theorem 58. *Let f be a Kasami Boolean function in n variables of degree t, where $t \geqslant 4$. It holds that*

(1) *if $t \leqslant n/2$, then $D_{a_1} \ldots D_{a_{t-3}} f(c)$ is not identically equal to 0, where $a_1, \ldots, a_{t-3}$ are arbitrary linearly independent elements of $\mathbb{F}_{2^n}$;*

(2) *if $t \leqslant (n/3) + 1$, then $D_{a_1} \ldots D_{a_{t-2}} f(c)$ is not identically equal to 0, where $a_1, \ldots, a_{t-2}$ are arbitrary linearly independent vectors of $\mathbb{F}_{2^n}$.*

A Boolean function in n variables is called *k-essentially dependent* if for each product of any k pairwise different variables there is a monomial in the algebraic normal form of f that contains this product.

Note that a function is k-essentially dependent if its multiple derivative with respect to $e_{i_1}, \ldots, e_{i_k}$ is not identically equal to 0 for all pairwise different $e_j = (0, \ldots, 0, 1, 0, \ldots, 0)$, where 1 is on position j, $j = 1, \ldots, n$. The following theorem holds [123]:

Theorem 59. *Let f be a Kasami Boolean function in n variables of degree $t > 3$, $n \geqslant 8$. It holds that*

(1) if $t = 4$, then f is 2-essentially dependent;

(2) if $t \leqslant n/2$, then f is $(t - 3)$-essentially dependent;

(3) if $t \leqslant (n/3) + 1$, then f is $(t - 2)$-essentially dependent.

Note that in the general case if $\deg(f) = t$, where $4 \leqslant t \leqslant n/2$, then the function f is not $(t - 1)$-essentially dependent.

For example suppose that $n = 12$, the field $\mathbb{F}_{2^n}$ is built with the irreducible polynomial $x^{12} + x^3 + 1$, $k = 5$, $d = 2^{2k} - 2^k + 1 = 993$, and $\lambda = x + 1$. Then the Kasami bent function $f(c) = \mathrm{tr}(\lambda c^d)$ has degree $\deg(f) = 6$ and is 4-essentially dependent but not 5-essentially dependent [123].

Thus, the order of essential dependence for Kasami Boolean functions of degree t equals $t - 3$ or $t - 2$.

9.6 CANTEAUT-LEANDER BENT FUNCTIONS (MF-1)

In 2004, Canteaut conjectured (on the basis of computer experiments) that $d = (2^k + 1)^2$ is a bent exponent when $n = 4k$ and k is odd. In 2006, Leander [230] proved this.

Theorem 60. *Let $d = (2^k + 1)^2$, where $n = 4k$ and k is odd. Then d is a bent exponent.*

Note that this monomial bent function belongs to the Maiorana-McFarland class (that is why this class is called MF-1). According to Theorem 10, bent functions from the MF-1 class are cubic.

In 2008, Charpin and Kyureghyan [80] showed that among cubic monomial bent functions only MF-1 and MF-2 bent functions (see the next section) belong to the Maiorana-McFarland class.

9.7 CANTEAUT-CHARPIN-KUYREGHYAN BENT FUNCTIONS (MF-2)

In 2008, Canteaut et al. [32] proved the following theorem:

Theorem 61. *The number $d = 2^{2k} + 2^k + 1$, where $n = 6k$, is a bent exponent.*

Moreover, in [32] necessary and sufficient conditions on a coefficient α for $f(c) = \mathrm{tr}(\alpha c^d)$ to be bent are given.

Note that these bent functions belong to the Maiorana-McFarland class (according to [226], we call them MF-2 bent functions). Again by Theorem 10 we see that bent functions from the MF-2 class are cubic.

9.8 NIHO EXPONENTS

The second series of bent functions contains functions of the form

$$f(c) = \mathrm{tr}(a_1 c^{d_1} + a_2 c^{d_2}) \tag{9.1}$$

for the appropriate elements $a_1, a_2 \in \mathbb{F}_{2^n}$ and exponents d_1, d_2. There are known examples of such functions with special exponents—*Niho exponents* of the form

$$d \equiv 2^i \quad \mathrm{mod}\ 2^{n/2} - 1.$$

Without loss of generality [112], let the first exponent be

$$d_1 = (2^{(n/2)} - 1)\frac{1}{2} + 1.$$

According to Dobbertin et al. [115], the following theorem holds:

Theorem 62. *If $d_2 = (2^{(n/2)} - 1)\lambda + 1$ and λ is equal to $1/6$, $1/4$, or 3, then there are elements $a_1, a_2 \in \mathbb{F}_{2^n}$ such that (9.1) is a bent function.*

Algorithmic questions for constructing bent functions with Niho exponents have been studied by Yang et al. [385]. Bent functions of the form

$$f(c) = \sum_{i=1}^{(n-1)/2} c_i\, \mathrm{tr}(c^{1+2^i})$$

have been studied by Khoo et al. [196, 197], Charpin et al. [81], Yu and Gong [390], and Hu and Feng [176], among others.

In 2006, Leander and Kholosha [231] proposed a new primary construction of bent functions consisting of a linear combination of 2^r Niho exponents.

Connections between Dillon bent functions, Niho bent functions, and oval polynomials can be found in a paper by Carlet and Mesnager [65] from 2011. See also the paper by Helleseth et al. [163].

In 2012, Budaghyan et al. [26] considered the Niho bent function consisting of 2^r exponents. The dual of such a function was found, and it was shown that this new bent function is not of the Niho type. All known univariate representations of Niho bent functions were analyzed for their relation to the completed Maiorana-McFarland class. In particular, it was proven [26] that two families do not belong to the completed class. The latter result gave a positive answer to an open problem of whether the class of bent functions introduced by Dillon [108] differs from the completed class of Maiorana-McFarland bent functions.

In 2013, upper bounds on algebraic immunity of Niho bent functions were studied by Gangopadhyay et al. [129].

9.9 GENERAL ALGEBRAIC APPROACH

Note that until now the algebraic constructions of bent functions have been casual: every time only functions of the special type are considered, and there is no general algebraic approach to the problem of constructing bent functions. This approach can be based on the fact that any Boolean function $f : \mathbb{F}_{2^n} \to \mathbb{F}_2$ can be represented in the trace form—that is,

$$f(c) = \mathrm{tr}\left(\sum_{j \in CS} a_j c^j\right) + \mathrm{tr}(a_{2^n-1} c^{2^n-1}),$$

for appropriate elements $a_j \in \mathbb{F}_{2^n}$, where CS is the set of representatives of cyclotomic classes modulo $2^n - 1$; see Theorem 6 in Chapter 1. An evolutional algorithm based on this idea was proposed by Yang et al. [385]. This work already been mentioned in Section 7.2 in connection with classification of bent functions in six and eight variables. Yang et al. [385] put forward several hypotheses about the trace form of a bent function in general. They suppose that in every class of extended affinely equivalent bent functions it is possible to choose a function with a trace form containing a small number of monomials. In opinion of Yang et al. [385], the monomials most probably in the trace form of a bent function are those for which d is a bent exponent. They based such a supposition on numerous computer experiments.

But there is still no general algebraic approach to the description of the class of bent functions.

9.10 OTHER CONSTRUCTIONS

Bent functions with special trace forms were considered in 2005 by Ma et al. [249].

Constructions of quadratic bent functions in polynomial form were proposed in 2014 by Li et al. [236].

CHAPTER 10

Bent Functions and Other Cryptographic Properties

INTRODUCTION

Since Boolean functions play a key role for constructing components of symmetric ciphers, their cryptographic properties are of great interest. In this chapter, we discuss the possibilities to combine "bentness" with other special characteristics of a Boolean function (high correlation immunity; resiliency; balancedness; high algebraic immunity, etc.).

10.1 CRYPTOGRAPHIC CRITERIA

Boolean functions that are used in cryptographic applications (e.g., in block or stream ciphers) should satisfy special conditions as far as it is necessary for guaranteeing the security of a cipher against different kinds of cryptanalysis. The cryptographic properties of Boolean functions consist of the following: high degree, balancedness, correlation immunity and r-resilience, high nonlinearity, algebraic immunity, etc.

In general, "cryptographic properties of Boolean functions" is a great (huge!) separate theme. In this book, we do not discuss when and why distinct cryptographic criteria have appeared: we deal with them in relation to bent functions only.

Cryptographic Boolean functions and their applications have been studied in the excellent and detailed books by Cusick and Stănică [96] (2009) and Logachev et al. [247] (2004, 2012). Very informative and helpful on this topic are the chapters by Carlet [46, 47] written for the monograph *Boolean Models and Methods in Mathematics, Computer Science, and Engineering* (2010). In Russian, there are also monographs by Tarannikov [351] (2011) and Pankratova [295] (2014). We highly recommend them to the interested reader.

Bent Functions
http://dx.doi.org/10.1016/B978-0-12-802318-1.00010-8

10.2 HIGH DEGREE AND BALANCEDNESS

Using Boolean functions of high degree in block ciphers leads to more complicated systems of equations (with key bits as unknown variables) describing the cipher and hence make cryptanalysis of the cipher more difficult. In pseudorandom generators, use of such functions helps to increase the linear complexity of the generated sequence. We recall here that the degree of a bent function in $n \geqslant 4$ variables is not more than $n/2$ (Theorem 16).

A Boolean function in n variables is called *balanced* if its weight equals 2^{n-1}. This means that the function takes the values 0 and 1 with the same probability, $1/2$. Clearly, balanced functions should be used in cryptosystems to resist the statistical methods of cryptanalysis. According to Theorem 22, the following theorem holds:

Theorem 63. *A bent function is never balanced.*

Balanced functions with the maximal possible nonlinearity are discussed in Section 17.2.

10.3 CORRELATION IMMUNITY AND RESILIENCY

A *subfunction of order k* of a Boolean function f in variables $x_1, \ldots, x_n$ is a function $f_{i_1,\ldots,i_k}^{a_1,\ldots,a_k}$ where each variable x_{i_j} is fixed by the value $a_{i_j}, j = 1, \ldots, k$, $0 < k \leqslant n$.

A Boolean function f is called *k-resilient* if any of its subfunctions of order not more than k is balanced.

A Boolean function f in n variables is called *correlation immune of order k* if the weight of any of its subfunction of order k equals $\mathrm{wt}(f)/2^k$. Note that a Boolean function f is k-resilient if and only if a function f is balanced and correlation immune of order k.

The property of resiliency is an extension of a weaker property—balancedness. Also there is an explicit connection between the properties of correlation immunity and resiliency.

It is well known that a Boolean function f is correlation immune of order k if and only if $W_f(y) = 0$ for all vectors y with $1 \leqslant \mathrm{wt}(y) \leqslant k$. Since all Walsh-Hadamard coefficients of a bent function are nonzero, the following theorem holds:

Theorem 64. *Every bent function in n variables is not a correlation-immune function. It is not a resilient function either.*

Minimal distances between bent and resilient functions were studied by Maity and Maitra [251, 252] in 2004 and 2009, and by Qu and Li [311] in 2009.

10.4 ALGEBRAIC IMMUNITY

Let f be a Boolean function. The minimum algebraic degree of a Boolean function g, $g \neq 0$, such that $f \cdot g = 0$ or $(f \oplus 1) \cdot g = 0$ is called the *algebraic immunity* of f, and is denoted by $\mathrm{AI}(f)$.

In ciphers, we should use Boolean functions with high algebraic immunity in order to avoid the application of algebraic cryptanalysis. Recall that algebraic attacks recover the secret key, or at least the initialization of the system, by solving a system of multivariate algebraic equations that describes a cipher.

It is well known that algebraic immunity of a Boolean function in n variables cannot be more than $\lceil n/2 \rceil$. Here $\lceil \cdot \rceil$ denotes the upper integer of a number—for example, $\lceil 3, 14 \rceil = 4$.

Algebraic immunity of bent functions from distinct classes has been studied recently.

In 2011, Gupta et al. [148] found a large subclass of Maiorana-McFarland bent functions (see Theorem 34) in n variables with a proven low level of algebraic immunity less than or equal to $\lceil n/4 \rceil + 2$.

Then in 2012 Wang and Tan [372] showed that the bound of Gupta et al. is not exact. They proved a more general theorem for a much larger subclass of Maiorana-McFarland functions, and found that its algebraic immunity cannot achieve the optimum value:

Theorem 65. *Let π be an arbitrary permutation on the set $\mathbb{F}_2^{n/2}$, and let h be an arbitrary Boolean function in $n/2$ variables. Then algebraic immunity of the Maiorana-McFarland bent function $f(x', x'') = \langle x', \pi(x'') \rangle \oplus h(x'')$ is not more than $\deg(\pi) + 1 + \lceil n/4 \rceil$.*

They found also an eight-variable Maiorana-McFarland function with algebraic immunity 4. Hence, this shows that there may be Maiorana-McFarland functions achieving the maximal algebraic immunity.

In fact, there is a very interesting question: Are there bent functions with the maximal possible algebraic immunity $n/2$ for any even n?

The answer to this question was studied by several authors. In 2011, Tu and Deng [367] for any even n constructed bent functions with algebraic immunity $n/2$ based on some combinatorial hypothesis—namely, the following conjecture:

Conjecture 1. Assume k is an integer, $k > 1$. For every integer m such that $0 \leqslant m \leqslant 2^k - 1$, we expand m as a binary string of length k, and denote the number of 1's in the string by wt(m). For any integer t, $0 < t < 2^k - 1$, let the set S_t consist of all pairs (a, b) such that a and b are integers, $0 \leqslant a, b < 2^k - 1$, $a + b = t \mod (2^k - 1)$, and wt(a) + wt(b) $\leqslant k - 1$. Then $|S_t| \leqslant 2^{k-1}$.

Tu and Deng [367] considered bent functions from the class $\mathcal{PS}_{ap}$ (Theorem 40) with the special balanced function g. Recall that by Theorem 40 if g is a balanced Boolean function in $n/2$ variables such that $g(0) = 0$, then the Boolean function f in n variables defined by $f(x, y) = g(xy^{2^{n/2}-2})$, for all $x, y \in \mathbb{F}_{2^{n/2}}$, is a bent function. Tu and Deng proved the following theorem:

Theorem 66. *Let g be the balanced Boolean function in $n/2$ variables given by its support:* supp(f) = $\{1 = \alpha^0, \alpha^1, \ldots, \alpha^{2^{(n/2)-1}-1}\}$*, where α is a primitive element of $\mathbb{F}_{2^{n/2}}$. Then if Conjecture 1 is true for $k = n/2$, the bent function $f(x, y) = g(xy^{2^{n/2}-2})$ has the maximal possible algebraic immunity $n/2$.*

In 2011, Tu and Deng were unable to prove Conjecture 1 mathematically, but they successfully developed a transfer-matrix algorithm, by which the conjecture was validated up to $k = 29$. Cusick et al. [97] proved that the conjecture is true for many cases of t, and they predicted that a general counting is difficult to obtain.

Later, in 2013, Deng [105] gave some arguments for the support of the conjecture.

In 2014, Filyuzin [121] studied algebraic immunity of a bent function from $\mathcal{PS}_{ap}$ constructed via the simplest balanced function. By "simplest" we mean an affine nonconstant Boolean function. The following theorem has been proven [121]:

Theorem 67. *Let g be an affine nonconstant Boolean function in $n/2$ variables. Then the bent function $f(x, y) = g(xy^{2^{n/2}-2})$ has algebraic immunity not more than $\lceil n/4 \rceil + 1$.*

Further investigations on algebraic immunity of Dillon's bent functions can be found in paper [122] by Filyuzin (2015).

In 2013, upper bounds on algebraic immunity of bent functions from some special classes such as $\mathcal{PS}_{ap}$ and Niho bent functions were studied by Gangopadhyay et al. [129].

10.5 VECTORIAL BENT FUNCTIONS, ALMOST BENT FUNCTIONS, AND ALMOST PERFECT NONLINEAR FUNCTIONS

Since the 1990s vectorial Boolean functions have been intensely studied in connection with their direct applications in cryptography. A function $F : \mathbb{F}_2^n \to \mathbb{F}_2^m$ is a *vectorial Boolean function* or (n, m)-*function*. In cryptographic terms, such a function is called an *S-box*.

We briefly discuss nonlinear properties of vectorial functions.

A *Walsh-Hadamard transform* of a vectorial (n, m)-function F is the following mapping $W_F^{\text{vect}} : \mathbb{F}_2^n \times \mathbb{F}_2^m \to \mathbb{Z}$, given by

$$W_F^{\text{vect}}(a, b) = \sum_{x \in \mathbb{F}_2^n} (-1)^{\langle a, x\rangle \oplus \langle b, F(x)\rangle} \quad \text{for any } a \in \mathbb{F}_2^n, b \in \mathbb{F}_2^m.$$

Nonlinearity of a (n, m)-function F is the minimum among nonlinearities of Boolean functions F_b in n variables, where $F_b(x) = \langle b, F(x)\rangle$ and $b \in \mathbb{F}_2^m$, $b \neq 0$. Then

$$N_F = \min_{b \in (\mathbb{F}_2^m)^*} \text{dist}(F_b, \mathcal{A}_n) = 2^{n-1} - \frac{1}{2} \max_{a \in \mathbb{Z}_2^n, b \in (\mathbb{F}_2^m)^*} |W_F^{\text{vect}}(a, b)|.$$

It is clear that as in the Boolean case

$$N_F \leqslant 2^{n-1} - 2^{(n/2)-1}. \tag{10.1}$$

A vectorial (n, m)-function is a *bent function* if N_F attains the maximal possible value—that is, every component function F_b, where $b \in (\mathbb{F}_2^m)^*$, is a (classical) bent function. Note that vectorial bent functions may exist only when n is even.

Theorem 68. *A vectorial (n, m)-function f is bent if and only if for any nonzero vector y the function $f(x) \oplus f(x \oplus y)$ is balanced—that is, takes every possible value (there are 2^m variants) exactly 2^{n-m} times.*

The following important fact was proven by Nyberg [288] in 1991:

Theorem 69. *A bent (n, m)-function exists if and only if $m \leqslant n/2$.*

It is easy to construct examples of vectorial bent functions using the vectorial form of the Maiorana–McFarland construction of Nyberg [288].

Theorem 70. *Let n be even and $m \leqslant n/2$. Let h be an arbitrary permutation on $\mathbb{F}_{2^{n/2}}$, and let g be an arbitrary $(n/2, m)$-function. Let $L : \mathbb{F}_{2^{n/2}} \to \mathbb{F}_2^{n/2}$ be a one-to-one linear or affine mapping. Then a vectorial function $f(x, y) = L(x \cdot h(y)) \oplus g(y)$ is a bent (n, m)-function.*

There are other constructions of bent functions that can be transferred on the vector case, see details in chapter of C. Carlet [47]. Other nonlinear characteristics of a vectorial function were discussed by V.V. Yashchenko in 1998 [387].

Since there are no bent (n, m)-functions when $m > n/2$, the bound (10.1) is not exact for this case. In 1971 Sidel'nikov [327] and independently Chabaud and Vaudenay in 1994 [76] proved the following fact.

Theorem 71. *Let $m \geqslant n - 1$. For any (n, m)-function f it is true*

$$N_f \leqslant 2^{n-1} - \frac{1}{2}\sqrt{3(2^n) - 2 - 2\frac{(2^n - 1)(2^{n-1} - 1)}{2^m - 1}}. \qquad (10.2)$$

If $n/2 < m < n - 1$ the bound better then (10.1) is still unknown.

The case $n = m$ is of special interest: For it the bound (10.2) has the form

$$N_f \leqslant 2^{n-1} - 2^{(n-1)/2}.$$

There is a large trend in cryptographic vectorial Boolean functions: to study almost bent (AB) and almost perfect nonlinear (APN) functions. Both of them can be considered as vectorial generalizations of bent functions. AB functions generalize bent functions in the classical sense, in terms of nonlinearity, (see Definition 1 in Section 2.3), whereas APN functions generalize bent functions in the sense of the Meier and Staffelbach approach [262] (see Definition 3 in Section 2.3). Note that APN functions are close to the *planar* (or *perfect nonlinear*) functions studied in projective geometry from 1968 by P. Dembowski and T. Ostrom.

A vectorial (n, n)-function F is called an *almost bent function* (AB function), if its nonlinearity N_f has the maximal possible value, $N_f = 2^{n-1} - 2^{(n-1)/2}$. A vectorial (n, n)-function F is called an *almost perfect nonlinear function* (APN function) if for every vector $a, b \in \mathbb{F}_2^n$, such that $a \neq 0$, the equation $F(x) \oplus F(x \oplus a) = b$ has no more than two solutions.

AB and APN functions are very widely studied and have several connections to classical bent functions. We do not consider them in this book, and we refer the interested reader to excellent surveys on this theme by Carlet [47] and Budaghyan [25], and also papers by Nyberg [289] and Tuzhilin [368]. Many new and still not widely known details about results in vectorial nonlinear Boolean functions can be found in a paper by Glukhov [138]. For example, it is not too well known that the first APN function was proposed by V.A. Bashev and studied by B.A. Egorov as long ago as 1968. It was an APN permutation $\tau : x \to x^{-1}$, such that $\tau(0) = 0$, for an odd n. For the even $n = 8$, this function (being differentially 4-uniform) was applied in the S-box of the Advanced Encryption Standard (AES).

Carlet [48] contributed to the problem of constructing APN functions from bent functions.

In 2014, vectorial bent functions with multiple trace terms were considered by Muratovic-Ribic et al. [281].

Akademgorodok, Novosibirsk

CHAPTER 11

Distances Between Bent Functions

INTRODUCTION

Hamming distances between bent functions are studied. In general, we follow the PhD thesis of N.A. Kolomeec published in 2014 and totally devoted to the topic of this chapter. It is shown that the minimal possible distance between two distinct bent functions in n variables is equal to $2^{n/2}$. Moreover, bent functions are at this distance if and only if they differ in all elements of some affine subspace of dimension $n/2$ and both functions are affine on it. It was proven by Kolomeec that if f is a bent function in n variables, then the number of bent functions at distance $2^{n/2}$ from it is not more than $2^{n/2} \prod_{i=1}^{n/2} (2^i + 1)$; this bound is achieved if and only if f is a quadratic bent function. Complete classification of all bent functions at distance $2^{n/2}$ from a quadratic bent function is given. Some new themes such as locally metrically nonequivalent bent functions are discussed. Finally, we consider the graph of minimal distances of bent functions and discuss a problem regarding its connectivity.

11.1 THE MINIMAL POSSIBLE DISTANCE BETWEEN BENT FUNCTIONS

The question about minimal distances between bent and resilient functions was considered by Maity and Maitra [251, 252] in 2004 and 2010, and Qu and Li [311] in 2009. Some metrical properties of self-dual bent functions were obtained by Carlet et al. [52]. Namely, they proved that the distance between a self-dual bent function and an anti-self-dual bent function, both in n variables, is 2^{n-1}.

Here we concentrate only on the following question: *What is the minimal Hamming distance between two distinct bent functions and when it is achieved?* We denote this minimal distance by $d_{\min}$—that is, $d_{\min} = \min\limits_{f,g \in \mathcal{B}_n; f \neq g} \operatorname{dist}(f, g)$.

One can prove the following theorem:

Theorem 72. $d_{\min} = 2^{n/2}$.

Bent Functions
http://dx.doi.org/10.1016/B978-0-12-802318-1.00011-X

Recall that a nonempty subset $L \subseteq \mathbb{F}_2^n$ is a *linear subspace* of $\mathbb{F}_2^n$ if the sum of any of its elements x and y belongs to L. A linear subspace with 2^k elements has *dimension k*. A shift of a linear subspace L on a some vector $z \in \mathbb{F}_2^n$—that is, the set $\{x \oplus z : x \in L\}$—is called an *affine subspace* of $\mathbb{F}_2^n$. Its *dimension* coincides with the dimension of L.

A Boolean function f in n variables is *affine on a set* $L \subseteq \mathbb{F}_2^n$ if there is a vector $a \in \mathbb{F}_2^n$ and a constant $b \in \mathbb{F}_2$ such that $f(x) = \langle a, x \rangle \oplus b$ for every $x \in L$.

In 1993, Carlet [36] proposed a construction of bent functions based on affine properties of Boolean functions. Although his construction was proposed without connections to metrics, it allows one to obtain bent functions at the minimal distance $2^{n/2}$ between one another. Namely [36], we have the following theorem:

Theorem 73. *Let f be a bent function in n variables. Let L be an affine subspace of $\mathbb{F}_2^n$ of dimension $n/2$. Let f be affine on L. Then a Boolean function $f \oplus \mathrm{Ind}_L$ is also a bent function in n variables.*

In 2009, Kolomeec and Pavlov [206] (see also [207]) proved that if two bent functions are at the minimal possible distance $2^{n/2}$, then one has to be obtained from the other via Carlet's construction. Let us formulate this result as a theorem:

Theorem 74. *Let f and g be Boolean functions in n variables. Let f be a bent function. Suppose that $\mathrm{dist}(f, g) = 2^{n/2}$. Then g is bent if and only if the set $\mathrm{supp}(f \oplus g)$ is an affine subspace and f is affine on it.*

Note that a bent function at distance $2^{n/2}$ does not exist for every bent function in n variables since not every bent function is normal and weakly normal; see Section 16.4.

11.2 CLASSIFICATION OF BENT FUNCTIONS AT THE MINIMAL DISTANCE FROM THE QUADRATIC BENT FUNCTION

It is well known that extended affine equivalence preserves Hamming distances between Boolean functions—that is,

$$\mathrm{dist}(f, g) = \mathrm{dist}(A(f), A(g)),$$

where $A(\cdot)$ is an arbitrary nondegenerate affine transformation of a Boolean function (i.e., a nondegenerate affine transformation of variables and adding an affine Boolean function; see Section 5.2).

According to Theorem 18, any quadratic bent function in n variables is equivalent to the function $q(x) = x_1 x_{(n/2)+1} \oplus x_2 x_{(n/2)+2} \oplus \cdots \oplus x_{n/2} x_n$. It is convenient for us to deal with a quadratic bent function of this form. Thus, it is interesting to study bent functions at the minimal distance from only one bent function q.

To classify all bent functions at distance $2^{n/2}$ from q let us introduce some notation. Let M be a binary $k \times n$ matrix in *reduced row echelon form* without zero rows. This means that

- the leading number (the first nonzero number from the left) of a nonzero row is always strictly to the right of the leading coefficient of the row above it;
- all entries in a column above a leading entry are zeroes;
- there are exactly k leading numbers.

Such a matrix is called also a *Gaussian-Jordan matrix* (without zero rows). So, M looks like

$$M = \begin{pmatrix} 1 & * & 0 & 0 & * & 0 & * & \ldots \\ 0 & 0 & 1 & 0 & * & 0 & * & \ldots \\ 0 & 0 & 0 & 1 & * & 0 & * & \ldots \\ 0 & 0 & 0 & 0 & 0 & 1 & * & \ldots \\ \ldots & & \ldots & & \ldots & & \ldots \end{pmatrix}.$$

Note that for every linear subspace $L \subseteq \mathbb{F}_2^n$ of dimension k there is a unique matrix M in the reduced row echelon form with rows being basic vectors. Then the matrix is usually called the *Gaussian-Jordan matrix of a subspace L*.

Denote by $\ell(M)$ the subset of positions of leading numbers of rows in a matrix M. For instance, in our example given above, $\ell(M) = \{1, 3, 4, 6, \ldots\}$.

In general, let the Gaussian-Jordan matrix have the form

$$M = \left(\begin{array}{c|c} A & B \\ \hline 0 & C \end{array} \right).$$

Let us call a Gaussian-Jordan $n/2 \times n$ matrix M *admissible of order t* if the following conditions hold:

- A is a Gaussian-Jordan $t \times n/2$ matrix.
- C is a Gaussian-Jordan $(n/2 - t) \times n/2$ matrix.
- Linear subspaces of $\mathbb{F}_2^{n/2}$ with Gaussian-Jordan matrices A and C are orthogonal—that is, $A \cdot C^T$ is the zero matrix.
- All columns of B with numbers from $\ell(C)$ are zero.

- A' and B' are matrices of size $t \times t$ obtained from A and B, respectively, after elimination of columns with numbers from $\ell(C)$. Then if $t > 1$,

$$
\begin{pmatrix}
a'^{(2)} & a'^{(1)} & 0 & 0 & \cdots & 0 \\
\cdots \\
a'^{(t)} & 0 & 0 & \cdots & 0 & a'^{(1)} \\
\cdots \\
0 & a'^{(3)} & a'^{(2)} & 0 & \cdots & 0 \\
\cdots \\
0 & a'^{(t)} & 0 & \cdots & 0 & a'^{(2)} \\
\cdots \\
0 & 0 & 0 & \cdots & a'^{(t)} & a'^{(t-1)}
\end{pmatrix}
\cdot
\begin{pmatrix}
b'^{(1)\,\mathrm{T}} \\
b'^{(2)\,\mathrm{T}} \\
\vdots \\
b'^{(t)\,\mathrm{T}}
\end{pmatrix}
= 0,
$$

where $a'^{(i)}$ and $b'^{(i)}$ are the ith rows of A' and B', respectively.

Kolomeec proved [208] the following theorem (see also [210]):

Theorem 75. *Let V be an affine subspace of $\mathbb{F}_2^n$ of dimension $n/2$. Then the bent function $q(x) = x_1 x_{(n/2)+1} \oplus x_2 x_{(n/2)+2} \oplus \cdots \oplus x_{n/2} x_n$ is affine on V if and only if V is a shift of a linear subspace with an admissible Gaussian-Jordan matrix.*

For instance, if $n = 4$, the bent function $x_1 x_2 \oplus x_3 x_4$ is affine on all linear subspaces (and all their shifts) with Gaussian-Jordan matrices

$$
\begin{pmatrix} 0 & 0 & 1 & 0 \\ 0 & 0 & 0 & 1 \end{pmatrix}, \quad
\begin{pmatrix} 1 & 0 & * & 0 \\ 0 & 0 & 0 & 1 \end{pmatrix}, \quad
\begin{pmatrix} 1 & 1 & 0 & * \\ 0 & 0 & 1 & 1 \end{pmatrix},
$$

$$
\begin{pmatrix} 0 & 1 & 0 & * \\ 0 & 0 & 1 & 0 \end{pmatrix}, \quad
\begin{pmatrix} 1 & 0 & * & 0 \\ 0 & 1 & 0 & * \end{pmatrix}, \quad
\begin{pmatrix} 1 & 0 & * & 1 \\ 0 & 1 & 1 & * \end{pmatrix}.
$$

Thus, by Theorems 74 and 75 in order to count the number of bent functions at a distance $2^{n/2}$ from the quadratic bent function one has to enumerate all the affine subspaces of dimension $n/2$ with admissible Gaussian-Jordan matrices. This is done in [208], where the following theorem is proven:

Theorem 76. *Any quadratic bent function in n variables has exactly $2^{n/2} \prod_{i=1}^{n/2} (2^i + 1)$ bent functions at the minimal possible distance $2^{n/2}$ from it.*

For instance, there are 60 bent functions at the minimal possible distance 4 from a bent function in four variables (since every such bent function is quadratic).

The number from Theorem 76 can be estimated as

$$2^{n(n+6)/8} < 2^{n/2} \cdot (2^1 + 1) \cdot \cdots \cdot (2^{n/2} + 1) < 3 \cdot 2^{n(n+6)/8}.$$

Note that asymptotically

$$2^{n/2} \cdot (2^1 + 1) \cdot \cdots \cdot (2^{n/2} + 1) \approx 2.38 \cdot 2^{n(n+6)/8}.$$

It is interesting that more than one-third of admissible Gaussian-Jordan matrices (that we use in Theorem 75) can be constructed in a very simple manner. Namely, all matrices of the form $\begin{pmatrix} E & T \end{pmatrix}$, where E is the identity matrix of size $n/2 \times n/2$ and matrix T is an arbitrary symmetric matrix of size $n/2 \times n/2$, are admissible Gaussian-Jordan matrices.

Note also that every bent function at the minimal distance $2^{n/2}$ from the quadratic bent function in n variables is equivalent to a McFarland bent function.

11.3 UPPER BOUND FOR THE NUMBER OF BENT FUNCTIONS AT THE MINIMAL DISTANCE FROM AN ARBITRARY BENT FUNCTION

Kolomeec studied bounds for the number of bent functions at the distance $2^{n/2}$ from an arbitrary bent function. In 2014, he proved [211] the following upper bound:

Theorem 77. *Let f be a bent function in n variables. Then the number of bent functions at distance $2^{n/2}$ from f is not more than*

$$2^{n/2} \prod_{i=1}^{n/2} (2^i + 1).$$

The bound is achieved if and only if f is a quadratic bent function.

The proof of Theorem 77 is based on the special affine property of quadratic bent functions that helps to separate them from the class of bent functions. We consider this property in more detail.

Kolomeec introduced the following definition. A Boolean function f in n variables is called *completely affinely decomposable of order k*, where $2 \leqslant k \leqslant n$, if f is affine on at least one affine subspace of $\mathbb{F}_2^n$ of dimension k, and if f is affine on an affine subspace $V \subseteq \mathbb{F}_2^n$ of dimension k, then it is affine on any of its shifts.

The following theorem is proven in [209]:

Theorem 78. *Let f be a Boolean function in n variables (here n is an arbitrary integer number). Then*

(1) *a function f is completely affinely decomposable of order* $2 \leqslant k \leqslant \lceil n/2 \rceil$ *if and only if f is affine or quadratic;*

(2) *a function f is completely affinely decomposable of order* $\lceil n/2 \rceil \leqslant k < n$ *and not of order* $k+1$ *if and only if f is extended affinely equivalent to the function* $x_1 x_2 \oplus x_3 x_4 \oplus \cdots \oplus x_{2n-2k-1} x_{2n-2k}$;

(3) *a function f is completely affinely decomposable of order n if and only if f is affine.*

Thus, only affine and quadratic Boolean functions can be completely affinely decomposable. Because the bent function $q(x) = x_1 x_{(n/2)+1} \oplus x_2 x_{(n/2)+2} \oplus \cdots \oplus x_{n/2} x_n$ exactly belongs to the class of completely affinely decomposable functions, this provides it with the special distance properties mentioned in Theorems 76 and 77.

11.4 BENT FUNCTIONS AT THE MINIMAL DISTANCE FROM A McFARLAND BENT FUNCTION

If f is a function from the McFarland class, then it is possible to get a lower bound for the number of bent functions at distance $2^{n/2}$ from it.

In 2012, Kolomeec [208] proved the following theorem:

Theorem 79. *Let f be a bent function in n variables extended affinely equivalent to a McFarland bent function. Then the number of bent functions at distance* $2^{n/2}$ *from it is not less than* $2^{n+1} - 2^{n/2}$.

11.5 LOCALLY METRICALLY EQUIVALENT BENT FUNCTIONS

First, let us consider some computational results [206] on metrical properties of bent functions for small n.

In Table 11.1 for each bent function f in n variables (representative of the equivalence class) one can see the number K_f of bent functions at the minimal distance $2^{n/2}$ from it, where $n = 4, 6, 8$. We mark data related to quadratic bent functions with bold type. Then it is easy to see (in accordance with Theorem 77) that the number of bent functions at the minimal distance from the quadratic bent function is significantly more than the other numbers in this column.

It is interesting that nonequivalent bent functions can have the same number of bent functions at the minimal distance. Let us introduce a new type of equivalence for bent functions—*local metrical equivalence.* We call two

Table 11.1 Number of bent functions at distance $2^{n/2}$ from a given bent function

n	Bent function f	Number K_f
4	**12 + 34**	**60**
6	123 + 245 + 346 + 14 + 26 + 34 + 35 + 36 + 45 + 46	376
6	123 + 245 + 12 + 14 + 26 + 35 + 45	440
6	123 + 14 + 25 + 36	568
6	**12 + 34 + 56**	**1080**
8	123 + 245 + 346 + 147 + 35 + 27 + 15 + 16 + 48	1392
8	123 + 245 + 346 + 35 + 26 + 25 + 17 + 48	2928
8	123 + 245 + 346 + 35 + 13 + 14 + 27 + 68	2928
8	123 + 245 + 346 + 35 + 26 + 25 + 12 + 13 + 14 + 78	2928
8	123 + 245 + 346 + 35 + 16 + 27 + 48	2928
8	127 + 347 + 567 + 14 + 36 + 25 + 45 + 78	4464
8	123 + 245 + 34 + 26 + 17 + 58	6000
8	123 + 245 + 13 + 15 + 26 + 34 + 78	6000
8	123 + 14 + 25 + 36 + 78	12 144
8	**12 + 34 + 56 + 78**	**36 720**

bent functions f and g in n variables *locally metrically equivalent* if the numbers of bent functions at distance $2^{n/2}$ from f and g coincide. It is clear that if two bent functions are extended affinely equivalent, then they are locally metrically equivalent.

Looking at Table 11.1, one can see that among bent functions of degree 3 or less in eight variables there are six classes of locally metrically nonequivalent bent functions, whereas there are 10 such classes with respect to "standard" (extended affine) equivalence. Note also that all nonweakly normal bent functions in n variables are locally metrically equivalent since each of them has zero bent functions at distance $2^{n/2}$.

11.6 THE GRAPH OF MINIMAL DISTANCES OF BENT FUNCTIONS

Let GB_n be the graph of bent functions in n variables as vertices with edges between functions that are at the minimal possible distance $2^{n/2}$ from each other. According to Theorem 77, the degrees of vertices in GB_n are not more than $2^{n/2} \prod_{i=1}^{n/2}(2^i + 1)$. It is clear that bent functions are locally metrically equivalent if and only if they have equal degrees in this graph.

Since for every even $n \geq 14$ there are nonweakly normal bent functions [33], graph GB_n is not connected if $n \geq 14$. It can be easily proven [213] that GB_n is connected for $n = 2, 4, 6$.

Thus, some open questions arise:

- Is the graph GB_n connected/disconnected if $8 \leqslant n \leqslant 12$?
- If GB'_n is the graph obtained from GB_n after elimination of all pendant vertices (corresponding to nonweakly normal bent functions), is GB'_n connected for all even $n \geqslant 2$?

In 2014, Kolomeec [205, 213] proved that the subgraph GM_n of GB_n induced by the vertices corresponding to McFarland and extended affinely equivalent to them bent functions is connected.

Note that Potapov [307] also studied metrical properties of bent functions in 2012. On the basis of the results obtained by Kasami and Tokura [192], he proved the following theorem:

Theorem 80. *Let f and g be bent functions in n variables such that $2^{n/2} \leqslant \mathrm{dist}(f,g) < 2^{n/2+1}$. Then $\mathrm{dist}(f,g)$ is equal to $2^{n/2+1} - 2^t$ for some t, where $0 \leqslant t \leqslant n/2$.*

Some other properties related to Hamming distances and bent functions will be considered in Chapter 12.

CHAPTER 12

Automorphisms of the Set of Bent Functions

INTRODUCTION

The theory of bent functions contains many unsolved problems; among them there is a question about the automorphism group of the set of all bent functions in n variables. In this chapter, we give a solution to this problem proposed by the author in 2010. First, we prove that for any nonaffine Boolean function f in n variables there is a bent function g in n variables such that the function $f \oplus g$ is not bent. This fact implies that affine Boolean functions are precisely all Boolean functions which are at the maximal possible distance from the class of bent functions. In other words, there is a *duality*, in some sense, between the definitions of bent functions and affine functions. As a corollary, we obtain that the set of bent functions and the set of affine functions have the same groups of automorphisms. This common group is a semidirect product of the general affine group $GA(n)$ and $\mathbb{Z}_2^{n+1}$.

12.1 PRELIMINARIES

Some attempts to determine the automorphism group of *a given* bent function were undertaken by Dempwolff [103] in 2006. Results were presented in terms of elementary Abelian Hadamard difference sets (equivalently, bent functions). Dempwolff described some series of such sets and computed their automorphism groups. For some of these sets the construction was based on the nonvanishing of the degree 1-cohomology of certain Chevalley groups in characteristic 2. Dempwolff classified bent functions f such that $\mathrm{Aut}(f)$ together with the translations from the underlying vector space induce a rank 3 group of automorphisms of the associated symmetric design. Computational aspects of the problem were also discussed in [103].

This chapter is devoted to determining the automorphism group of the *whole set of bent functions* in n variables.

Bent Functions
http://dx.doi.org/10.1016/B978-0-12-802318-1.00012-1

Let us recall several notions. We remember that a bent function can be defined as follows (Definition 3 in Chapter 2). A Boolean function g is *bent* if and only if for all nonzero y

$$\sum_{x \in \mathbb{F}_2^n} (-1)^{g(x) \oplus g(x \oplus y)} = 0.$$

Let A be a binary nonsingular $n \times n$ matrix over $\mathbb{F}_2$, let b and c be binary vectors of length n, and let d be a constant. It is well known that any transform $g(x) \to g(Ax \oplus b) \oplus \langle c, x \rangle \oplus d$ maps a bent function to a bent function. Moreover any such mapping is *isometric*—that is, it preserves Hamming distances between functions. Then a natural question arises: *Are there other isometric mappings of Boolean functions into themselves that preserve the class of bent functions without changes?* In this chapter, we show that there are *no* other mappings possessing such a property. In this chapter, we follow [360].

12.2 SHIFTS OF THE CLASS OF BENT FUNCTIONS

Consider the vectors with fixed values of some $n - k$ coordinates; let the remaining coordinates be arbitrary. The set of all such vectors is called the *facet of dimension k* of the space $\mathbb{F}_2^n$. For example, the set $\Gamma = \{(x', x'') : x'' = a\}$ is a facet of dimension $n/2$, where $x', x'' \in \mathbb{F}_2^n$ and a is a fixed vector.

Branches

Now let us prove the basic and most complicated fact of this chapter:

Theorem 81. *For each nonaffine Boolean function f in n variables there is a bent function g in n variables such that $f \oplus g$ is not bent.*

Proof. Prove it from the contrary. Suppose that for a fixed nonaffine f and for an arbitrary bent function g the function $f \oplus g$ is bent.

The idea of the proof is as follows. First, we show that some special sum has to be equal to 0 for any bent function; see sum (12.1). Then we find a bent function g' in the McFarland class which is a counterexample and for which this equality does not hold. The bent function g' will be obtained from a specially chosen bent function g by inversion of its values on some facet Γ of dimension $n/2$. The key condition for the possibility to choose such a facet is that for some nonzero y the set $D = \mathrm{supp}(f(x) \oplus f(x \oplus y))$ is a proper subset of the space $\mathbb{F}_2^n$. This is possible if and only if f is nonaffine. It is convenient to divide the proof into several steps:

• *Step 1. Preliminaries.* Since f is nonaffine, there is a nonzero vector y such that $D_y f(x) = f(x) \oplus f(x \oplus y)$ is not a constant function. Hence, the set $D = \mathrm{supp}(D_y f)$ is nonempty and does not coincide with $\mathbb{F}_2^n$. Without loss of generality, suppose $y = \mathbf{1}$, where $\mathbf{1} = (1, \ldots, 1)$. We can do this since it is always possible to change the function $f(x)$ to an arbitrary function $f(Ax)$, where $A \cdot \mathbf{1} = y$.

• *Step 2. The equality for a bent function.* Let g be a bent function in n variables. Then $f \oplus g$ is also bent, and by definition

$$\sum_x (-1)^{g(x) \oplus g(x \oplus y)} = 0$$

and

$$\sum_x (-1)^{g(x) \oplus f(x) \oplus g(x \oplus y) \oplus f(x \oplus y)} = 0.$$

Subtracting the second equality from the first one, we get

$$\sum_x (-1)^{g(x) \oplus g(x \oplus y)} (1 - (-1)^{f(x) \oplus f(x \oplus y)}) = 0.$$

Thus, for an arbitrary bent function g,

$$\sum_{x \in D} (-1)^{g(x) \oplus g(x \oplus y)} = 0. \qquad (12.1)$$

• *Step 3. A choice of the set.* Since D is nonempty and does not coincide with $\mathbb{F}_2^n$, there is a facet Γ in the Boolean cube $\mathbb{F}_2^n$ such that $\dim \Gamma = n/2$ and both sets $\Gamma \cap D$ and $\Gamma \cap (\mathbb{F}_2^n \setminus D)$ are nonempty—that is,

$$0 < m < |\Gamma| = 2^{n/2}, \tag{12.2}$$

where $m = |\Gamma \cap D|$. Indeed, we can construct this facet Γ in such a way that it includes some vector $u \notin D$ and one of the vectors v or $v \oplus y$, where $v \in D$. Note that one of the distances $d(u,v)$ or $d(u, v \oplus y)$ is not more than $n/2$.

$|(\Gamma \oplus y) \cap D| = m$ as far as $D \oplus y = D$. Facets Γ and $\Gamma \oplus y$ do not intersect according to the choice of y (recall that $y = 1$). Then the set D can be decomposed as follows:

$$D = (\Gamma \cap D) \cup ((\Gamma \oplus y) \cap D) \cup (D \setminus (\Gamma \cup (\Gamma \oplus y))). \tag{12.3}$$

We use this decomposition further. Without loss of generality, consider $\Gamma = \{(x', x'') : x'' = a\}$ for some $a \in \mathbb{F}_2^{n/2}$. Let

$$\Gamma \cap D = \{(b^{(1)}, a), (b^{(2)}, a), \ldots, (b^{(m)}, a)\}$$

for the corresponding vectors $b^{(i)}$.

- *Step 4. A special subset of bent functions.* Now we need the special subclass of bent functions—namely, the Maiorana-McFarland class; see Theorem 34 in Chapter 8. Let G be the subclass of bent functions of the McFarland type $g(x', x'') = \langle x', \pi(x'') \rangle$, where $\pi(x'') = Ax''$ and A is a nonsingular matrix. Let us show that there is a function $g \in G$ such that

$$S = \sum_{x \in \Gamma \cap D} (-1)^{g(x) \oplus g(x \oplus y)} \neq 0.$$

In fact, for each $g \in G$ we have $g(x) \oplus g(x \oplus y) = \langle x', Ax'' \rangle \oplus \langle x' \oplus y', Ax'' \oplus Ay'' \rangle = \langle x', Ay'' \rangle \oplus \langle y', Ax'' \oplus Ay'' \rangle$, since $\pi(x'' \oplus y'') = A(x'' \oplus y'') = Ax'' \oplus Ay''$, where $y = (y', y'')$. Then,

$$S = (-1)^{\gamma} \sum_{i=1}^{m} (-1)^{\langle b^{(i)}, Ay'' \rangle},$$

where $\gamma = \langle y', Aa \oplus Ay'' \rangle$ is a constant depending on the concrete choice of A. Since y'' is nonzero (we recall that $y'' = 1$) and A is an arbitrary nonsingular matrix, the vector $z = Ay''$ can be an arbitrary nonzero vector of length $n/2$. So, let us show that there is a nonzero vector z such that

$$\sum_{i=1}^{m}(-1)^{\langle b^{(i)},z\rangle} \tag{12.4}$$

is not equal to 0.

• Step 5. A search for the vector z. Assume the contrary. Let for all nonzero z sum (12.4) be equal to 0. Consider the binary matrix M of size $(n/2) \times m$ with columns $b^{(1)},\ldots,b^{(m)}$. Vector z defines the linear combination of rows of M: we take row i in combination if and only if $z_i = 1$. Then sum (12.4) is the difference between the number of 0's and the number of 1's in this linear combination. By our assumption, every nonzero linear combination of rows of M should contain half of the 0's and half of the 1's. Hence, the matrix M should have the following form (up to the permutation of columns):

$$\begin{pmatrix} 0\ldots0 & 0\ldots0 & 0\ldots0 & 0\ldots0 & 1\ldots1 & 1\ldots1 & 1\ldots1 & 1\ldots1 \\ 0\ldots0 & 0\ldots0 & 1\ldots1 & 1\ldots1 & 0\ldots0 & 0\ldots0 & 1\ldots1 & 1\ldots1 \\ 0\ldots0 & 1\ldots1 & 0\ldots0 & 1\ldots1 & 0\ldots0 & 1\ldots1 & 0\ldots0 & 1\ldots1 \\ \cdots & \cdots & \cdots & \cdots & \cdots & \cdots & \cdots & \cdots \\ \underbrace{}_{m/8} & \underbrace{}_{m/8} & \underbrace{}_{m/8} & \underbrace{}_{m/8} & \underbrace{}_{m/8} & \underbrace{}_{m/8} & \underbrace{}_{m/8} & \underbrace{}_{m/8} \end{pmatrix}.$$

From here we obtain that the number of columns in M should be not less than $2^{n/2}$, where $n/2$ is the number of rows. Thus, it should be $m \geqslant 2^{n/2}$. But this contradicts the choice of facet Γ—namely, inequality (12.2).

Therefore, there is a vector z such that (12.4) is nonzero. Let us fix this vector z.

• Step 6. Construction of the function—counterexample. Let A be a nonsingular matrix such that $Ay'' = z$. Let permutation π be defined by $\pi(x'') = Ax''$. According to the choice of vector z,

$$S = \sum_{x \in \Gamma \cap D} (-1)^{g(x) \oplus g(x \oplus \gamma)} \neq 0 \tag{12.5}$$

for the function $g(x',x'') = \langle x', \pi(x'') \rangle$. Now define the function $g'(x',x'') = g(x',x'') \oplus t(x'')$ of McFarland type by the rule

$$t(x'') = \begin{cases} 1, & \text{if } x'' = a, \\ 0, & \text{in other cases.} \end{cases}$$

Since g is bent, g' is bent too.

By decomposition (12.3) we get

$$\sum_{x\in D}(-1)^{g'(x)\oplus g'(x\oplus y)} = \left(\sum_{x\in \Gamma\cap D}(-1)^{g(x)\oplus 1\oplus g(x\oplus y)\oplus 0}\right)$$
$$+\left(\sum_{x\in (\Gamma\oplus y)\cap D}(-1)^{g(x)\oplus 0\oplus g(x\oplus y)\oplus 1}\right)$$
$$+\left(\sum_{x\in (D\setminus(\Gamma\cup(\Gamma\oplus y)))}(-1)^{g(x)\oplus g(x\oplus y)}\right).$$

Thus,

$$\sum_{x\in D}(-1)^{g'(x)\oplus g'(x\oplus y)} = \left(\sum_{x\in (D\setminus(\Gamma\cup(\Gamma\oplus y)))}(-1)^{g(x)\oplus g(x\oplus y)}\right) - 2S$$
$$= \sum_{x\in D}(-1)^{g(x)\oplus g(x\oplus y)} - 4S.$$

Therefore by (12.1) and (12.5) we obtain

$$\sum_{x\in D}(-1)^{g'(x)\oplus g'(x\oplus y)} \neq 0.$$

We see now that equality (12.1) cannot be satisfied for bent functions g and g' simultaneously. This contradiction proves the theorem.

$\square$

12.3 DUALITY BETWEEN DEFINITIONS OF BENT AND AFFINE FUNCTIONS

Now we study the definitions of bent and affine functions.

For an even n the class of bent functions is

$$\mathcal{B}_n = \{f : \mathrm{dist}(f, \mathcal{A}_n) = N_{\max}\},$$

where $N_{\max} = 2^{n-1} - 2^{(n/2)-1}$. Is it possible to *invert* this definition? In other words, is it true that $\mathcal{A}_n$ is the set of all Boolean functions that are at the maximal distance (say, $N'_{\max}$) from $\mathcal{B}_n$? Here we see there are several questions. What is $N'_{\max}$? What is the connection between $\mathcal{A}_n$ and $\mathcal{A}'_n$, where

$$\mathcal{A}'_n = \{f : \mathrm{dist}(f, \mathcal{B}_n) = N'_{\max}\}?$$

We have proven that this inversion of definitions is right. The "duality theorem" [364] holds:

Theorem 82. *A Boolean function in n variables is*

(1) *a bent function if and only if it is at the maximal possible distance $N_{\max}$ from the set of all affine functions;*

(2) *an affine function if and only if it is at the maximal possible distance $N_{\max}$ from the set of all bent functions.*

Proof. We have to prove the second statement only. Let us prove two sufficient facts for it.

Fact 1. $N'_{\max} = 2^{n-1} - 2^{(n/2)-1}$. By definition $N'_{\max} = \max_f \min_{g \in \mathcal{B}_n} \mathrm{dist}(f, g)$. Note that $\mathrm{dist}(f, g) = 2^{n-1} - \frac{1}{2}W_{f \oplus g}(0)$. Hence,

$$N'_{\max} = 2^{n-1} - \frac{1}{2}\min_f \max_{g \in \mathcal{B}_n}|W_{f \oplus g}(0)|.$$

Let us fix an arbitrary bent function g' in n variables. Since the class $\mathcal{B}_n$ is closed under addition of affine functions, every function $g' \oplus \ell_a$ is bent, where $\ell_a(x) = \langle a, x \rangle$. Note that

$$W_{f \oplus g' \oplus \ell_a}(0) = W_{f \oplus g'}(a).$$

Now it is obvious that

$$\max_{g \in \mathcal{B}_n}|W_{f \oplus g}(0)| \geqslant \max_{\substack{g \in \mathcal{B}_n, g = g' \oplus \ell_a \\ \text{for some } a}}|W_{f \oplus g}(0)| = \max_a|W_{f \oplus g'}(a)|.$$

From Parseval's equality for the Boolean function $f \oplus g'$ it follows that $\max_a|W_{f \oplus g'}(a)| \geqslant 2^{n/2}$. Now we obtain $N'_{\max} \leqslant 2^{n-1} - 2^{(n/2)-1}$. On the other hand, the distance $2^{n-1} - 2^{(n/2)-1}$ to the class $\mathcal{B}_n$ is attained, for example, for an affine function f.

Fact 2. $\mathcal{A}_n = \mathcal{A}'_n$.

It is obvious that $\mathcal{A}_n \subseteq \mathcal{A}'_n$. Suppose that there is a function $f \in \mathcal{A}'_n \setminus \mathcal{A}_n$. Then by Theorem 81 we can find a bent function g such that $f \oplus g$ is not a bent function. In other words, there is a vector a such that $|W_{f \oplus g}(a)| > 2^{n/2}$. Consider a bent function $g'(x) = g(x) \oplus \langle a, x \rangle$. $W_{f \oplus g'}(0) = W_{f \oplus g}(a)$, and by the equality $\mathrm{dist}(f, \mathcal{B}_n) = 2^{n-1} - \frac{1}{2}\max_{g \in \mathcal{B}_n}|W_{f \oplus g}(0)|$ we get $\mathrm{dist}(f, \mathcal{B}_n) < N'_{\max}$. This contradicts the choice of f. Thus, $\mathcal{A}_n = \mathcal{A}'_n$.

The theorem follows from these two facts.   □

12.4 AUTOMORPHISMS OF THE SET OF BENT FUNCTIONS

Recall that a mapping φ of the set of all Boolean functions in n variables into itself is *isometric* if it preserves Hamming distances between functions—that is, $\mathrm{dist}(\varphi(f), \varphi(g)) = \mathrm{dist}(f, g)$. From Theorems 17, 81, and 82, we have the following theorem:

Theorem 83. *Every isometric mapping of the set of all Boolean functions into itself that transforms bent functions into bent functions is a combination of an affine transform of coordinates and an affine shift—that is, it has the form*

$$g(x) \rightarrow g(Ax \oplus b) \oplus \langle c, x \rangle \oplus d.$$

It is known that any such a mapping can be uniquely represented as

$$g(x) \rightarrow g(s(x)) \oplus f(x), \tag{12.6}$$

where $s : \mathbb{F}_2^n \rightarrow \mathbb{F}_2^n$ is an arbitrary substitution, and f is an arbitrary function in n variables.

The *group of automorphisms* of a subset of Boolean functions $\mathcal{M}$ is the group of isometric mappings of the set of all Boolean functions into itself preserving the set $\mathcal{M}$. We denote this group by $\mathrm{Aut}(\mathcal{M})$.

Recall that the *general affine group* $\mathrm{GA}(n)$ consists of all mappings of the form $g(x) \rightarrow g(Ax \oplus b)$, where A is a nonsingular matrix and b is an arbitrary vector. The following assertion is true:

Theorem 84. *The group $\mathrm{Aut}(\mathcal{A}_n)$ is equal to the semidirect product of the general affine group $\mathrm{GA}(n)$ and $\mathbb{Z}_2^{n+1}$.*

Indeed, for any automorphism (12.6) the shift by a function f can be defined only for an affine function (since the image of the null function is also an affine function). The set of all affine functions of n variables forms a group isomorphic to $\mathbb{Z}_2^{n+1}$. It remains for us to note that, as is well known, each permutation s must belong to the group $\mathrm{GA}(n)$; see, for example, [261].

Theorem 85. *The following equalities hold:*

$$\mathrm{Aut}(\mathcal{B}_n) = \mathrm{Aut}(\mathcal{A}_n) = \mathrm{GA}(n) \ltimes \mathbb{Z}_2^{n+1}.$$

Proof. It is obvious that $\mathrm{Aut}(\mathcal{A}_n) \subseteq \mathrm{Aut}(\mathcal{B}_n)$. Indeed, for any mapping $\varphi \in \mathrm{Aut}(\mathcal{A}_n)$ and any bent function g,

$$\mathrm{dist}(g, \mathcal{A}_n) = \mathrm{dist}(\varphi(g), \varphi(\mathcal{A}_n)) = \mathrm{dist}(\varphi(g), \mathcal{A}_n).$$

Therefore, any bent function is transformed by φ to some other bent function.

Similarly, $\text{Aut}(\mathcal{B}_n) \subseteq \text{Aut}(\mathcal{A}_n)$. Namely, for an arbitrary automorphism $\psi \in \text{Aut}(\mathcal{B}_n)$ and any affine function f we have $\text{dist}(f, \mathcal{B}_n) = \text{dist}(\psi(f),$ $\psi(\mathcal{B}_n)) = \text{dist}(\psi(f), \mathcal{B}_n)$. In view of Theorem 82, it follows that $\psi(\mathcal{A}_n) = \mathcal{A}_n$.

We see that $\text{Aut}(\mathcal{B}_n) = \text{Aut}(\mathcal{A}_n)$. The form of this group follows from Theorem 84.   □

Thus, if mapping (12.6) transforms the class of bent functions into itself, then it is of the form $g(x) \rightarrow g(Ax \oplus b) \oplus \langle c, x \rangle \oplus d$. Recall that the bent functions derived from one another by such a mapping are called *equivalent* (or *extended affinely equivalent*). It follows from the results obtained that a more general approach to equivalence of bent functions (on the basis of isometric mappings) than that does not exist.

12.5 METRICALLY REGULAR SETS

It is very interesting to study the following general question.

Let $\mathcal{A}$ be a subset of $\mathbb{F}_2^n$. Let $\mathcal{B}$ be the set of all binary vectors from $\mathbb{F}_2^n$ that are at the maximal possible distance from the set $\mathcal{A}$. Now let $\mathcal{A}'$ be the set of all vectors that are at the maximal possible distance from $\mathcal{B}$.

We call a set $\mathcal{A}$ *metrically regular* if $\mathcal{A} = \mathcal{A}'$. Note that $\mathcal{A}$ and $\mathcal{B}$ are metrically regular (or not metrically regular) simultaneously.

For example, the set $\mathcal{A}$ consisting of all-0 and all-1 vectors only is metrically regular. In this case, the set $\mathcal{B}$ contains all vectors of Hamming weight $n/2$ (if n is even) or $(n-1)/2$, $(n+1)/2$ (if n is odd).

We call a subset of Boolean functions *metrically regular* if the set of corresponding vectors of values is metrically regular. In this chapter, we have proven that

- the set $\mathcal{A}_n$ of all affine Boolean functions in n variables (n is even) is metrically regular;
- the set $\mathcal{B}_n$ of all bent functions in n variables (n is even) is metrically regular.

The following open problem can be mentioned: *What kind of sets in $\mathbb{F}_2^n$ are metrically regular? Is it possible to obtain a classification of them?* With respect to Boolean functions, one can ask: What kind of classes of Boolean functions in n variables are metrically regular? For instance, is the set of all Boolean functions of degree less than or equal to k when $k \geqslant 2$ metrically regular?

In 2015 Oblaukhov [292] described metrical regular sets that can be represented as special subspaces of the Boolean cube.

CHAPTER 13

Bounds on the Number of Bent Functions

INTRODUCTION

One of the most important questions in the theory of bent functions is how many there are. If n is equal to 2, 4, 6, or 8, the number of bent functions in n variables is 8, 896, about $2^{32.3}$, and about $2^{106.29}$, respectively. If $n > 10$, the exact number of bent functions in n variables is still unknown. Moreover, there is a large gap between the lower $(2^{2^{(n/2)+\log_2(n-2)-1}})$ and upper $(2^{2^{n-1}+\frac{1}{2}\binom{n}{n/2}})$ bounds for this number. There have been several improvements of these bounds, but they are not too big. It is interesting that an analogous problem of the large gap between lower and upper bounds occurs for other combinatorial objects (e.g., error-correcting codes). To find the asymptotic value of the number of all bent functions is a long-standing hard problem closely connected to the problem of enumeration of Hadamard matrices (unsolved since 1893). In this chapter we collect results about bounds on the number of bent functions in n variables and discuss open problems in this area.

13.1 PRELIMINARIES

Two simple ideas are usually exploited to estimate of the number of some special combinatorial objects (functions, configurations, graphs, etc.): use some *restrictions* in the properties of these objects that distinguish them from the others in order to find an upper bound; use the known *constructions* of the objects in order to get a lower bound.

The same ideas are applicable in the case of bent functions. The known trivial upper bound for $|\mathcal{B}_n|$ is based on the degree restriction: an arbitrary bent function in n variables has degree not more than $n/2$ (Theorem 86). The simple lower bound is based on the Maiorana-McFarland construction. As far as constructions of bent functions are divided into two groups—*direct* and *iterative* (see Chapter 8)—lower bounds can also be divided in this way. We distinguish *direct* lower bounds, which can be written as formulas

Bent Functions
http://dx.doi.org/10.1016/B978-0-12-802318-1.00013-3

depending on n, and *iterative* lower bounds, which are usually implicit and "connect" the number of bent functions in n variables with such numbers in fewer variables. The best direct and iterative lower bounds for the number of bent functions are given in Theorems 89 and 91, respectively.

13.2 THE NUMBER OF BENT FUNCTIONS FOR SMALL n

First we present the known bounds on the number of bent functions in n variables when n is small.

The best bounds are collected in Table 13.1.

Remark. Let us explain how we count the bounds.

The bound on minterms (Theorem 90) does not work starting with 12 variables, since the exact number of bent functions in 10 variables is unknown. In fact, we count $|\mathcal{B}_{12}| = 6|\mathcal{B}_{10}|^2 - 8|\mathcal{B}_{10}|$ using the current best lower bound ($2^{262.12}$) and current best upper bound (2^{638}) for $|\mathcal{B}_{10}|$. Then we see that $|\mathcal{B}_{12}| \geq 6 \cdot 2^{2 \cdot 262.12} - 8 \cdot 2^{638} = 3 \cdot 2^{525.24} - 2^{641}$. Since this number is negative, the bound does not work in this case: we put "$-$" in Table 13.1.

We have a similar situation with the bound on bent iterative functions (Theorem 91). It gives a zero lower bound starting with 14 variables, since the bound in the case of 12 variables is too low. In detail, $|\mathcal{B}_{14}| \geq |\mathcal{B}_{12}|^4/|X_{12}|$. We use the lower bound $2^{410.64}$ for $|\mathcal{B}_{12}|$ and the upper bound 2^{2510} for $|X_{12}|$. Thus, $|\mathcal{B}_{14}| \geq 2^{4 \cdot 410.64 - 2510}$. This number is between 0 and 1 and so, we put "0" in Table 13.1.

We are grateful to Agievich [10], who has counted the concrete values of his bound (Theorem 89) when n is small.

13.3 UPPER BOUNDS

The trivial upper bound follows from the fact that the degree of a bent function is not more than $n/2$ (Theorem 16). We have the following theorem:

Theorem 86. $|\mathcal{B}_n| \leq 2^{1 + \binom{n}{1} + \binom{n}{2} + \cdots + \binom{n}{n/2}} = 2^{2^{n-1} + \frac{1}{2}\binom{n}{n/2}}$.

Carlet and Klapper [63] improved this bound a little in 2002:

Theorem 87. *Let $n \geq 6$ and $\varepsilon = \frac{1}{2^{O(\sqrt{2^n/n})}}$. Then,*

$$|\mathcal{B}_n| \leq 2^{2^{n-1} + \frac{1}{2}\binom{n}{n/2} - 2^{n/2} + (n/2) + 1}(1 + \varepsilon) + 2^{2^{n-1} - \frac{1}{2}\binom{n}{n/2}},$$

$$n = 2$$

Lower bounds:	
Minterm construction (Theorem 90)	—
Bent iterative functions (Theorem 91)	—
McFarland functions (Theorem 88)	8
Agievich's algorithm (Theorem 89)	8
Exact value of $|\mathcal{B}_2|$	8
Upper bounds:	
Trivial: based on degree restriction (Theorem 86)	8

$$n = 4$$

Lower bounds:	
Minterm construction (Theorem 90)	320
McFarland functions (Theorem 88)	384
Bent iterative functions (Theorem 91)	512
Agievich's algorithm (Theorem 89)	896
Exact value of $|\mathcal{B}_4|$	896
Upper bounds:	
Trivial: based on degree restriction (Theorem 86)	2048

$$n = 6$$

Lower bounds:	
Minterm construction (Theorem 90)	$2^{22.19}$
McFarland functions (Theorem 88)	$2^{23.3}$
Bent iterative functions (Theorem 91)	$2^{28.26}$
Agievich's algorithm (Theorem 89)	$2^{29.65}$
Exact value of $|\mathcal{B}_6|$ is $5\,425\,430\,528$	$\simeq 2^{32.3}$
Upper bounds:	
Proven by Carlet and Klapper [63]	2^{38}
Trivial: based on degree restriction (Theorem 86)	2^{42}

$$n = 8$$

Lower bounds:	
McFarland functions (Theorem 88)	$2^{60.25}$
Minterm construction (Theorem 90)	$2^{67.25}$
Agievich's algorithm (Theorem 89)	$2^{70.4}$
Bent iterative functions (Theorem 91)	$2^{87.34}$
Exact value of $|\mathcal{B}_8|$ is in Section 7.3	$\simeq 2^{106.29}$
Upper bounds:	
Proven by Langevin et al. [229]	$2^{129.2}$
Trivial: based on degree restriction (Theorem 86)	2^{163}

$$n = 10$$

Lower bounds:	
McFarland functions (Theorem 88)	$2^{149.66}$
Agievich's algorithm (Theorem 89)	$2^{163.03}$
Minterm's construction (Theorem 90)	$2^{215.16}$
Bent iterative functions (Theorems 91 and 95)	$2^{262.16}$
Exact value of $|\mathcal{B}_{10}|$	Unknown
Upper bounds:	
Trivial: based on degree restriction (Theorem 86)	2^{638}

$$n = 12$$

Lower bounds:	
Minterm construction (Theorem 90)	—
McFarland functions (Theorem 88)	2^{360}
Agievich's algorithm (Theorem 89)	$2^{376.46}$
Bent iterative functions (Theorem 91)	$2^{410.64}$
Exact value of $|\mathcal{B}_{12}|$	Unknown
Upper bounds:	
Trivial: based on degree restriction (Theorem 86)	2^{2510}

$$n = 14$$

Lower bounds:	
Minterm construction (Theorem 90)	—
Bent iterative functions (Theorem 91)	0
McFarland functions (Theorem 88)	$2^{844.16}$
Agievich's algorithm (Theorem 89)	$2^{863.67}$
Exact value of $|\mathcal{B}_{14}|$	Unknown
Upper bounds:	
Trivial: based on degree restriction (Theorem 86)	2^{9908}

$$n = 16$$

Lower bounds:	
Minterm construction (Theorem 90)	—
Bent iterative functions (Theorem 91)	0
McFarland functions (Theorem 88)	2^{1940}
Agievich's algorithm (Theorem 89)	$2^{1962.53}$
Exact value of $|\mathcal{B}_{16}|$	Unknown
Upper bounds:	
Trivial: based on degree restriction (Theorem 86)	$2^{39\,203}$

Table 13.1 Best bounds on the number of bent functions in n variables

n	Best bounds
2	$|\mathcal{B}_2| = 8$
4	$|\mathcal{B}_4| = 896$
6	$|\mathcal{B}_6| = 5\,425\,430\,528 \simeq 2^{32.3}$
	Counted by Preneel [310], and later by Agievich [7]; see also [264]
8	$|\mathcal{B}_8| = 2^9 \times 193\,887\,869\,660\,028\,067\,003\,488\,010\,240 \simeq 2^{106.29}$
	Counted by Langevin and Leander [228]
10	$2^{262.16} \leqslant |\mathcal{B}_{10}| \leqslant 2^{638}$
	Lower bound is from Theorems 91 and 95 [363]. Upper bound is from Theorem 86
12	$2^{410.64} \leqslant |\mathcal{B}_{12}| \leqslant 2^{2510}$
	Lower bound is from Theorem 91 [363]. Upper bound is from Theorem 86
14	$2^{863.67} \leqslant |\mathcal{B}_{14}| \leqslant 2^{9908}$
	Lower bound is from Theorem 89 [7]. Upper bound is from Theorem 86
16	$2^{1962.53} \leqslant |\mathcal{B}_{16}| \leqslant 2^{39\,203}$
	Lower bound is from Theorem 89 [7]. Upper bound is from Theorem 86

where

$$\varepsilon = 2^{-\left(\binom{n-1}{d-1}-\binom{n-1-d}{d-1}-1\right)}, \quad d \text{ is positive, } d \leqslant n.$$

But this upper bound is still close to the trivial upper bound.

A computational upper bound for the number of bent functions was discussed by Wang and Zhang [371] (2004). They introduced a bent matrix connected to a bent function. On the basis of its properties Wang and Zhang [371] obtained a computational upper bound that was claimed to be "the best possible" after consideration of examples. But there is insufficient detail in the paper to verify this.

Upper and lower bounds of the number of bent functions were also discussed by Zhao [397] in 2007.

13.4 DIRECT LOWER BOUNDS

Here we mention the direct lower bound that follows from the Maiorana-McFarland construction (see Theorem 34).

Theorem 88. $|\mathcal{B}_n| \geqslant 2^{2^{n/2}}(2^{n/2})!.$

Indeed, there are $(2^{n/2})!$ ways to choose a permutation on $\mathbb{F}_2^{n/2}$, and $2^{2^{n/2}}$ ways for an arbitrary Boolean function in $n/2$ variables to be taken. Asymptotically, the bound from Theorem 88 is as follows:

$$\left(\frac{2^{(n/2)+1}}{e}\right)^{2^{n/2}}\sqrt{2^{(n/2)+1}\pi}, \quad \text{or roughly,} \quad 2^{2^{(n/2)+\log_2(n-2)-1}}.$$

In 2000 Agievich [7] proposed an algorithm for generating distinct bent functions using bent squares (see Section 6.7). Let $\mathcal{BS}_n$ ("*bent squares*") be the set of all bent functions in n variables generated by that algorithm. Then $\mathcal{BS}_n$ contains all Maiorana-McFarland functions (as bent functions with the simplest bent squares; see Theorem 35) and some other bent functions with bent squares that are more complicated. For each n the size of $\mathcal{BS}_n$ can be computed directly. This estimation gives the following direct lower bound; see [7]. It is the best known direct lower bound for the number of bent functions.

Theorem 89. $|\mathcal{B}_n| \geqslant |\mathcal{BS}_n| = 2^{2^{n/2}}(2^{n/2})! + \sum_{d=2}^{2^{n-1}} N_d$, *where*

$$N_d = \left(\left(\frac{(2^n-1)(2^{n-1})!}{(2^{n-1}-d)!}\right)^2 \cdot 2^{4d} \cdot \sum_{i=0}^{d}\frac{(-1)^i}{i!} - \delta_n(d)\right) \cdot (2^n - 2d)! \cdot 2^{2^n - 2d}.$$

Here

$$\delta_n(d) = (d/2)! \cdot 2^{9d/2} \cdot \left(\frac{4}{3}s_n(d)r_n(d) - \frac{4}{9}r_n^2(d)\right)$$

if d is even and 0 otherwise, and

$$s_n(d) = (2^n - 1) \cdot \frac{(2^{n-1})!}{2^{d/2}(2^{n-1}-d)!(d/2)!},$$

$$r_n(d) = (2^n - 1)(2^{n-1} - 1)\binom{2^{n-2}}{d/2}.$$

It is difficult to characterize an asymptotic behavior of this bound, but for a small number of variables, we see that Agievich's bound is indeed bigger than MacFarland's bound.

13.5 ITERATIVE LOWER BOUNDS

As already mentioned, iterative lower bounds are based on iterative constructions of bent functions. Let us briefly review them with respect to this question about the number.

Iterative constructions for bent functions have been investigated by many authors. We recall only some of them here.

Recall that the first iterative construction was given by Rothaus [317, 318]. It says that a Boolean function $f(x, y) = g(x) \oplus h(y)$ is bent if and only if functions g and h are bent (Theorem 32). From here one can get a bound $|\mathcal{B}_n| \geqslant \max_{m+k=n} |\mathcal{B}_m||\mathcal{B}_k|$.

The next construction was obtained by Rothaus [317] and Dillon [108]. Let f', f'', and f''' be bent functions in n variables such that their sum is bent again. Then $g(a_1, a_2, x) = f'(x)f''(x) \oplus f'(x)f'''(x) \oplus f''(x)f'''(x) \oplus a_1 f'(x) \oplus a_1 f''(x) \oplus a_2 f'(x) \oplus a_2 f'''(x) \oplus a_1 a_2$ is a bent function in $n + 2$ variables (Theorem 43). But to obtain the lower bound on $|\mathcal{B}_n|$ from this construction seems to be rather difficult. In general, it is not clear when for distinct collections $\{f_1', f_1'', f_1'''\}$ and $\{f_2', f_2'', f_2'''\}$ one can obtain distinct bent functions g.

Another construction was introduced by Carlet [43]. Suppose that f', f'', and f''' are bent functions in n variables such that their sum (denote it by s) is bent. Moreover, let $\tilde{s} = \tilde{f'} \oplus \tilde{f''} \oplus \tilde{f'''}$. Then the function $g(x) = f'(x)f''(x) \oplus f'(x)f'''(x) \oplus f''(x)f'''(x)$ is bent in n variables (Theorem 44). In this case one can try to find an equality containing $|\mathcal{B}_n|$ and a certain function of $|\mathcal{B}_n|^3$. But we guess that checking conditions (when s is bent and when $\tilde{s} = \tilde{f'} \oplus \tilde{f''} \oplus \tilde{f'''}$) is difficult enough.

Climent et al. [91] proposed iterative constructions of bent functions in $n + 2$ variables from bent functions in n variables using minterms (see Theorems 46 and 47 in Chapter 8). Counting of distinct bent functions obtained via these constructions gives some new lower bounds for the number of all bent functions.

From Theorem 46, the next theorem follows [91]:

Theorem 90. $|\mathcal{B}_{n+2}| \geqslant 6|\mathcal{B}_n|^2 - 8|\mathcal{B}_n|$.

According to Climent et al. [93], Theorems 46 and 47 give the following bound: $|\mathcal{B}_{n+2}| \geqslant 6|\mathcal{B}_n|^2 + 2^{n+2}(2^n - 3)|\mathcal{B}_n|$. But it is not better than the bound of Theorem 90 when $n \geqslant 6$.

The iterative constructions of Korsakova [214] (Theorem 45) give linear in $|\mathcal{B}_n|$ iterative lower bounds, and that is why they are not too interesting.

An extensive study of the restrictions of bent functions to affine subspaces was proposed by Canteaut and Charpin [31]. In particular, they have established that restrictions of a bent function f to a subspace V of codimension 2 and to its cosets are bent if and only if the derivative of $\tilde{f}$ with respect to $V^{\perp}$ is constant equal to 1. This last result can be interpreted as an iterative construction for bent functions (Theorem 48).

The following iterative lower bound is based on this construction. It was proven by the author [363] in 2011. For now it is the best iterative lower bound for the number of bent functions.

Theorem 91. *For any even* $n \geqslant 2$, $|\mathcal{B}_{n+2}| \geqslant |\mathcal{BI}_{n+2}| \geqslant |\mathcal{B}_n|^4/|X_n|$, *where X_n is the set of all Boolean functions in n variables that can be represented as the sum of two distinct bent functions.*

In the next section, the proof of this theorem is given.

13.6 LOWER BOUND FROM THE BENT ITERATIVE FUNCTIONS

First, let us present a way to calculate the number of bent iterative functions in terms of special finite sets.

Recall the following notion. Let a Boolean function g in $n + 2$ variables be defined as

$$g(00, x) = f_0(x), \quad g(01, x) = f_1(x), \quad (13.1)$$
$$g(10, x) = f_2(x), \quad g(11, x) = f_3(x),$$

where f_0, f_1, f_2, and f_3 are Boolean functions in n variables. Note that for distinct ordered collections $\{f_0, f_1, f_2, f_3\}$ we always obtain distinct functions g. From [31] it follows that if functions f_0, f_1, and f_2 are bent functions in n variables, then function g defined by (13.1) is a bent function in $n + 2$ variables if and only if f_3 is a bent function in n variables and $\widetilde{f_0} \oplus \widetilde{f_1} \oplus \widetilde{f_2} \oplus \widetilde{f_3} = 1$.

Bent function g obtained in this way is called *a bent iterative function*. Recall that $\mathcal{BI}_n$ denotes the class of all bent iterative functions in n variables. For more details, see Theorem 48 in Section 8.8.

Now we prove the following theorem:

Theorem 92. *For any even* $n \geqslant 4$

$$|\mathcal{BI}_n| = \sum_{f' \in \mathcal{B}_{n-2}} \sum_{f'' \in \mathcal{B}_{n-2}} | (\mathcal{B}_{n-2} \oplus f') \cap (\mathcal{B}_{n-2} \oplus f'') |.$$

Proof. Let us study in how many ways one can construct a bent iterative function g in n variables. We can do this as follows. First, take an arbitrary ordered pair of two bent functions f_0, f_1 in $n - 2$ variables. These functions may coincide or may not coincide. The number of all distinct such pairs is $|\mathcal{B}_{n-2}|^2$. Then take a suitable bent function f_2 in $n - 2$ variables. We call bent function f_2 *suitable* for f_0, f_1 if function $\widetilde{f_0} \oplus \widetilde{f_1} \oplus \widetilde{f_2}$ is bent. It is clear that according to Theorem 48 for any suitable bent function f_2 one can construct a bent iterative function g by determining f_3 from the equality

$\tilde{f_3} = \tilde{f_0} \oplus \tilde{f_1} \oplus \tilde{f_2} \oplus 1$. So, if $k(f_0, f_1)$ is the number of suitable bent functions f_2 for given bent functions f_0 and f_1, then

$$|\mathcal{BI}_n| = \sum_{f_0 \in \mathcal{B}_{n-2}} \sum_{f_1 \in \mathcal{B}_{n-2}} k(f_0, f_1). \qquad (13.2)$$

Indeed, any bent iterative function g can be obtained in the way presented. Note that g is uniquely determined by the ordered triple f_0, f_1, f_2.

Now let us study numbers $k(f_0, f_1)$. Let $\mathcal{B}(f_0, f_1)$ be the set of all suitable bent functions in $n - 2$ variables for f_0, f_1. So, $|\mathcal{B}(f_0, f_1)| = k(f_0, f_1)$. Show that the set $\mathcal{B}(f_0, f_1)$ is in one-to-one correspondence with $(\mathcal{B}_{n-2} \oplus \tilde{f_0}) \cap (\mathcal{B}_{n-2} \oplus \tilde{f_1})$.

Define a map $\phi : \mathcal{B}(f_0, f_1) \to (\mathcal{B}_{n-2} \oplus \tilde{f_0}) \cap (\mathcal{B}_{n-2} \oplus \tilde{f_1})$ by the rule $\phi(f_2) = \tilde{f_0} \oplus \tilde{f_2}$, for all $f_2 \in \mathcal{B}(f_0, f_1)$. First check that ϕ is defined correctly. Since f_2 is suitable, there is a bent function h in $n - 2$ variables such that $h = \tilde{f_0} \oplus \tilde{f_1} \oplus \tilde{f_2}$. Then function $s = \tilde{f_0} \oplus \tilde{f_2}$ belongs to the set $\mathcal{B}_{n-2} \oplus \tilde{f_0}$ and also belongs to the set $\mathcal{B}_{n-2} \oplus \tilde{f_1}$ as far as $s = \tilde{f_1} \oplus h$. Thus, ϕ is defined correctly.

Prove that ϕ is a bijective mapping. It is easy to see that if $f_2 \neq f_2'$, then $\phi(f_2) \neq \phi(f_2')$. Let s be a function from $(\mathcal{B}_{n-2} \oplus \tilde{f_0}) \cap (\mathcal{B}_{n-2} \oplus \tilde{f_1})$. Then bent function f_2 defined by $\tilde{f_2} = s \oplus \tilde{f_0}$ is suitable for f_0, f_1 (since function $\tilde{f_0} \oplus \tilde{f_1} \oplus \tilde{f_2} = \tilde{f_1} \oplus s$ is bent), and $\phi(f_2) = s$. Thus, ϕ is bijective.

So, it is proven that $k(f_0, f_1) = |(\mathcal{B}_{n-2} \oplus \tilde{f_0}) \cap (\mathcal{B}_{n-2} \oplus \tilde{f_1})|$. Now replace $k(f_0, f_1)$ by this expression in formula (13.2) and change variables f_0, f_1 to variables $f' = \tilde{f_0}, f'' = \tilde{f_1}$. In such a way we get the statement of the theorem. $\qquad \square$

As an example, construct all bent iterative functions for $n = 4$. The total number of them is $|\mathcal{BI}_4| = \sum_{f' \in \mathcal{B}_2} \sum_{f'' \in \mathcal{B}_2} |(\mathcal{B}_2 \oplus f') \cap (\mathcal{B}_2 \oplus f'')|$. Recall that $\mathcal{B}_2$ consists of all functions with an odd number of nonzero values, $|\mathcal{B}_2| = 8$. Let us present their vectors of values: $(0001), (0010), (0100), (1000), (1110), (1101), (1011), (0111)$. Note that any set $\mathcal{B}_2 \oplus f'$ is the set of all Boolean functions with an even number of 1's. So, $|(\mathcal{B}_2 \oplus f') \cap (\mathcal{B}_2 \oplus f'')| = 8$ for any bent functions f', f''. This means that any bent function in two variables is suitable for all fixed bent functions f_0, f_1. Then, by Theorem 92 we get $|\mathcal{BI}_4| = 8 \cdot 8 \cdot 8 = 512$. Recall that $\mathcal{B}_4$ consists of 896 functions.

It is known that if a function f' is bent, then a function $f' \oplus \ell$ is bent too for every affine Boolean function ℓ. That is why any set $(\mathcal{B}_{n-2} \oplus f') \cap (\mathcal{B}_{n-2} \oplus f'')$ contains at least all affine functions in $n - 2$ variables: $\ell = (f' \oplus \ell) \oplus f' = (f'' \oplus \ell) \oplus f''$. The number of these affine functions is 2^{n-1}, and hence for any bent functions f', f'' in $n - 2$ variables we get

$2^{n-1} \leqslant | (\mathcal{B}_{n-2} \oplus f') \cap (\mathcal{B}_{n-2} \oplus f'') | \leqslant |\mathcal{B}_{n-2}|$. We note that for two bent functions f', f'' in $n-2$ variables $| (\mathcal{B}_{n-2} \oplus f') \cap (\mathcal{B}_{n-2} \oplus f'') | = |\mathcal{B}_{n-2}|$ if and only if $f' \oplus f''$ is an affine function as follows from Theorem 81.

Then from Theorem 92 we get the following corollary:

Corollary 1. $2^{n-1}|\mathcal{B}_{n-2}|^2 < |\mathcal{BI}_n| < |\mathcal{B}_{n-2}|^3$ *for any even $n \geqslant 4$.*

Indeed, it is enough to note that there are bent functions f', f'' for which $| (\mathcal{B}_{n-2} \oplus f') \cap (\mathcal{B}_{n-2} \oplus f'') | \leqslant |\mathcal{B}_{n-2}|$ holds and there are bent functions f', f'' for which it does not hold.

Corollary 2. $|\mathcal{BI}_n| > 2^{2^{(n/2)+2}-n-3}$ *for any even $n \geqslant 4$.*

Proof. From Corollary 1 it follows that $|\mathcal{BI}_n| > 2^{n-1}|\mathcal{B}_{n-2}|^2$. Since $\mathcal{BI}_{n-2}$ is a subset of $\mathcal{B}_{n-2}$, we have $|\mathcal{B}_{n-2}| > |\mathcal{BI}_{n-2}|$ and hence $|\mathcal{BI}_n| > 2^{n-1}|\mathcal{BI}_{n-2}|^2$. Applying Corollary 1 and inequality $|\mathcal{B}_{n-4}| > |\mathcal{BI}_{n-4}|$, we obtain $|\mathcal{BI}_n| > 2^{n-1} \cdot 2^{(n-3) \cdot 2} \cdot |\mathcal{BI}_{n-4}|^{2^2}$. We continue in this way and obtain

$$|\mathcal{BI}_n| > 2^{n-1} \cdot 2^{(n-3) \cdot 2} \cdot 2^{(n-5) \cdot 2^2} \cdot \ldots \cdot 2^{3 \cdot 2^{(n/2)-2}} |\mathcal{B}_2|^{2^{(n/2)-1}}.$$

Then, by substitution of $|\mathcal{B}_2| = 2^3$ we get $|\mathcal{BI}_n| > 2^{3 \cdot 2^{(n/2)-1}} \cdot 2^d$, where

$$d = (n-1) + (n-3) \cdot 2 + (n-5) \cdot 2^2 + \cdots + (n - (n-3)) \cdot 2^{(n/2)-2}.$$

One can see that

$$d = n \left(\sum_{i=0}^{(n/2)-2} 2^i \right) - \sum_{i=0}^{(n/2)-2} (2i+1)2^i = (n-1)(2^{(n/2)-1} - 1)$$

$$- 2 \sum_{i=0}^{(n/2)-2} i \cdot 2^i.$$

Using the combinatorial formula $\sum_{i=0}^{k} i \cdot 2^i = (k-1)2^{k+1} + 2$, we get $d = 5 \cdot 2^{(n/2)-1} - n - 3$. Hence, $|\mathcal{BI}_n| > 2^{2^{(n/2)+2}-n-3}$. $\qquad\square$

It is interesting to find better bounds on $|\mathcal{BI}_n|$ and to clarify if this number is closer to $|\mathcal{B}_{n-2}|^2$ or to $|\mathcal{B}_{n-2}|^3$. The answer to this question has a direct application to the problem of the lower bound for $|\mathcal{B}_n|$.

Define the following set $X_n = \{ f \oplus h : f, h \in \mathcal{B}_n \}$ and consider the system $\{C_f : f \in \mathcal{B}_n\}$ of its subsets defined as $C_f = \mathcal{B}_n \oplus f$. So,

$$X_n = \bigcup_{f \in \mathcal{B}_n} C_f.$$

One can prove that $|\mathcal{B}_n| > \sqrt{2|X_n|}$ for any even $n \geqslant 2$.

Let ψ be an element of X_n. We call the number of subsets C_f that cover ψ the *multiplicity* of ψ and denote it by $m(\psi)$. Note that if ψ is covered by C_f, then it is covered by any set $C_{f'}$, where f' is obtained from f by adding an affine function. It is clear that $\sum_{\psi \in X_n} m(\psi) = |\mathcal{B}_n|^2$.

Theorem 93. $|\mathcal{B}_{n+2}| \geqslant |\mathcal{BI}_{n+2}| = \sum_{\psi \in X_n} m(\psi)^2$ *for any even $n \geqslant 2$.*

Proof. The statement follows from Theorem 92. Indeed, let us fix an arbitrary function ψ from X_n. It is covered exactly by $m(\psi)$ sets—say, $C_{f^1}, C_{f^2}, \ldots, C_{f^{m(\psi)}}$. Now let pair (f', f'') run through all the ordered pairs of bent functions in n variables. Then function ψ is covered by set $(\mathcal{B}_n + f') \cap (\mathcal{B}_n + f'')$ if and only if functions f' and f'' both belong to the set $\{f^1, f^2, \ldots, f^{m(\psi)}\}$. The number of such ordered pairs is $m(\psi)^2$. Then, by Theorem 92 we get the required formula. $\qquad\square$

So, to evaluate $|\mathcal{BI}_{n+2}|$ (and then $|\mathcal{B}_{n+2}|$) we have to study the set X_n and the distribution of multiplicities for its elements.

Now let us prove Theorem 91. Recall that it says that *for any even $n \geqslant 2$,* $|\mathcal{B}_n| \geqslant |\mathcal{BI}_{n+2}| \geqslant |\mathcal{B}_n|^4/|X_n|$.

Proof. Since bent iterative functions form a subclass of the class of all bent functions, let us prove the second inequality only.

Previously we mentioned the equality $\sum_{\psi \in X_n} m(\psi) = |\mathcal{B}_n|^2$. Note that the minimal value of the sum $\sum_{\psi \in X_n} m(\psi)^2$ is reachable if and only if all the multiplicities are the same—that is, if and only if $m(\psi) = |\mathcal{B}_n|^2/|X_n|$ for all $\psi \in X_n$. Then, by Theorem 93 we have

$$|\mathcal{B}_{n+2}| \geqslant |\mathcal{BI}_{n+2}| = \sum_{\psi \in X_n} m(\psi)^2 \geqslant |X_n| \cdot \left(\frac{|\mathcal{B}_n|^2}{|X_n|} \right)^2 = \frac{|\mathcal{B}_n|^4}{|X_n|}.$$

$\qquad\square$

Corollary 3. *The average value of the square of the multiplicity in X_n is not less than $|\mathcal{B}_n|^4/|X_n|^2$.*

Since the degree of a bent function in n variables, $n \geqslant 4$, is not more than $n/2$ (Theorem 16), the set X_n can include only functions of degree less than or equal to $n/2$. Therefore, the following corollary holds:

Corollary 4. *For any even $n \geqslant 4$,*

$$|\mathcal{B}_{n+2}| \geqslant |\mathcal{BI}_{n+2}| \geqslant \frac{|\mathcal{B}_n|^4}{2^{1+n+\binom{n}{2}+\cdots+\binom{n}{n/2}}} = \frac{|\mathcal{B}_n|^4}{2^{2^{n-1}+\frac{1}{2}\binom{n}{n/2}}}.$$

To find the exact number of bent iterative functions, one has to find the distribution of multiplicities in X_n. Thus, we come to a new problem statement.

Open problem: bent sum decomposition. *What Boolean functions can be represented as the sum of two bent functions in n variables? How many such representations does a Boolean function admit?*

We suppose that the answers to these questions can be given in terms of probability theory. This problem is discussed in detail in Chapter 14.

13.7 TESTING OF THE LOWER BOUND FOR SMALL *n*

Now we study a behavior of the iterative lower bound from Section 13.6 (Theorems 91 and 93) when n is small. As will be shown for small values of n the bound from Theorem 91 is "almost exact" for evaluating the number $|\mathcal{BI}_n|$. Hence, in Section 13.8 several hypotheses will be proposed.

For $n = 2$ the set X_2 consists of all Boolean functions with an even number of nonzero values, $|X_2| = 8$. The multiplicities of all functions from X_2 are maximal and equal to 8. Thus, by Theorem 93 we have $|\mathcal{BI}_4| = 8 \cdot 8^2 = 512$. Recall that $\mathcal{B}_4$ consists of 896 functions.

For $n = 4$ the set X_4 consists of all Boolean functions of degree not more than 2, $|X_4| = 2^{11} = 2048$. All affine functions in X_4 (their number is 2^5) have maximal multiplicities equal to 896. All the others have multiplicities equal to 384. Thus, $|\mathcal{BI}_6| = 32 \cdot 896^2 + 2016 \cdot 384^2 = 77 \cdot 2^{22} = 322\,961\,408 \approx 2^{28.26}$. Note that via the Maiorana-McFarland construction [260] (with a fixed division of variables into halves) it is possible to obtain only $2^8(2^3)! = 315 \cdot 2^{15} = 10\,321\,920 \approx 2^{23.3}$ bent functions. The total number of bent functions in six variables is about $2^{32.3}$; see [362] for details.

If $n = 6$ the set X_6 again is the set of all Boolean functions of degree less than or equal to 3, $|X_6| = 2^{42}$. It was checked via an exhaustive search. We used an Intel Core i7 3.0 GHz processor 256 GB of RAM. The program worked for 14 days with full loading of RAM.

Now we describe the probabilistic investigation. We applied Monte Carlo methods for enumerating the sum $\sum_{\psi \in X_6} m(\psi)^2$, which is equal to $|\mathcal{BI}_8|$ by Theorem 93. We took at random $N = 346\,981 \approx 2^{18.4}$ Boolean functions in six variables of degree not more than 3 without linear parts in algebraic normal forms (by the linear part we mean all algebraic normal form items of degree less than or equal to 1). When checking the multiplicities

Table 13.2 Distribution of multiplicities in a sample of $N = 346\,981$ Boolean functions in six variables

i	m_i	n_i	i	m_i	n_i	i	m_i	n_i
1	$26\,880 \cdot 2^7$	102	11	$54\,784 \cdot 2^7$	67\,960	21	$82\,176 \cdot 2^7$	179
2	$33\,024 \cdot 2^7$	28	12	$56\,064 \cdot 2^7$	240	22	$83\,200 \cdot 2^7$	265
3	$36\,096 \cdot 2^7$	327	13	$56\,832 \cdot 2^7$	8559	23	$86\,784 \cdot 2^7$	109
4	$46\,464 \cdot 2^7$	38\,946	14	$57\,088 \cdot 2^7$	2130	24	$91\,392 \cdot 2^7$	56
5	$46\,848 \cdot 2^7$	12\,641	15	$57\,600 \cdot 2^7$	4	25	$119\,616 \cdot 2^7$	238
6	$47\,616 \cdot 2^7$	67\,687	16	$62\,208 \cdot 2^7$	596	26	$121\,600 \cdot 2^7$	42
7	$48\,896 \cdot 2^7$	6327	17	$63\,360 \cdot 2^7$	6073	27	$172\,800 \cdot 2^7$	22
8	$50\,496 \cdot 2^7$	36\,417	18	$65\,088 \cdot 2^7$	11\,019	28	$237\,312 \cdot 2^7$	6
9	$51\,968 \cdot 2^7$	12\,655	19	$66\,048 \cdot 2^7$	4272	29	$272\,640 \cdot 2^7$	15
10	$53\,952 \cdot 2^7$	67\,906	20	$80\,640 \cdot 2^7$	2159	30	$1\,521\,408 \cdot 2^7$	1

for these N functions, we found that there are only 30 distinct values of them. In Table 13.2 one can see the distribution of multiplicities for the functions taken. The number of Boolean functions with multiplicity m_i is denoted by n_i.

We then count the *sample average value* Q for the square of the multiplicity,

$$Q = \left(\sum_{i=1}^{30} n_i \cdot m_i^2 \right) / N = 45\,508\,981\,169\,513.30 \approx 2^{45.37}.$$

Since $|X_6| = 2^{42}$, we obtain the estimation

$$|\mathcal{BI}_8| \approx Q \cdot |X_6| = 200\,150\,615\,856\,476\,000\,000\,000\,000 \approx 2^{87.37}.$$

Now we evaluate the error of our approximation of $|\mathcal{BI}_8|$. In Monte Carlo methods one has to choose the *reliability* v of an approximation, $0 < v < 1$. The closer to 1 v is, the higher the reliability. Then the approximate upper bound for the error of our estimation can be obtained by the known formula

$$\delta = t_v S / \sqrt{N}$$

(see, e.g., [140]), where S is the *corrected standard deviation* for our approximation and t_v is the standard parameter determined by v. We get the value S by the known formula

$$S = \sqrt{\left(\sum_{i=1}^{30} n_i (m_i^2 - Q)^2 \right) / (N - 1)}.$$

So, $S = 65\,975\,029\,301\,812.10$. Now let $\nu = 0.999$. The corresponding standard parameter is $t_\nu = 3.291$ (see [140] for details). Then the approximate upper bound for the error is

$$\delta = 368\,599\,402\,514.14.$$

This means that with probability 0.999 the average value of the square of the multiplicity in the set X_6 is in the interval $(Q - \delta; Q + \delta)$. And hence we have proven the following theorem:

Theorem 94. *With probability* 0.999,

$$2^{87.36} < |\mathcal{BI}_8| < 2^{87.38}.$$

Note that according to Langevin and Leander [228], the total number of bent functions in eight variables is about $2^{106.29}$. Note also that the number of Maiorana–McFarland bent functions in eight variables is just about $2^{60.25}$.

We see that Q is close to the lower bound $|B_n|^4/|X_n|^2$ from Corollary 3.

For $n = 6$ this lower bound is $44\,793\,743\,175\,843.84 \approx 2^{45.348}$. So, from Theorem 91 it follows that $|\mathcal{BI}_8| > 197\,004\,891\,331\,091\,000\,000\,000\,000 \approx 2^{87.35}$. Since $|\mathcal{B}_8| \simeq 2^{106.29}$, by Theorem 91 we have the following theorem [363]:

Theorem 95. $|\mathcal{B}_{10}| > |\mathcal{BI}_{10}| > 830\,602\,255\,559\,379 \cdot 10^{64} > 2^{262.16}$.

13.8 ASYMPTOTIC PROBLEM AND HYPOTHESES

One of the main open problems in bent functions is to find the asymptotic value of the number of them. It is very difficult to make any progress in this area. Indeed, for $n \geqslant 10$ the number of bent functions in n variables is unknown. And there is a large gap between the lower $(2^{2^{(n/2)+\log(n-2)-1}})$ and upper $(2^{2^{n-1}+\frac{1}{2}\binom{n}{n/2}})$ bounds for this number. There are several improvements of these bounds [7, 63, 371], but they are not too significant with respect to $\log\log|\mathcal{B}_n|$. Here by log we mean $\log_2$. In this section several hypotheses based on the results obtained and the calculations performed are formulated.

Results for X_2, X_4, and X_6 lead us to the following strong hypothesis:

Hypothesis 1. *Every Boolean function in n variables of degree not more than $n/2$ can be represented as the sum of two bent functions in n variables (n is even, $n \geqslant 2$).*

So, we suppose that X_n is as large as possible—that is, $|X_n| = 2^{2^{n-1}+\frac{1}{2}\binom{n}{n/2}}$. In other words, we conjecture that for any Boolean function f in n variables there is a bent function g in n variables such that $f \oplus g$ is bent. Note that this hypothesis has similarity to the previously obtained fact (see Theorem 81 in Chapter 12): for any nonaffine Boolean function f in n variables there is a bent function g in n variables such that $f \oplus g$ is not bent.

If Hypothesis 1 is correct, then by inequality $|\mathcal{B}_n| > \sqrt{2|X_n|}$ one can prove the following hypothesis:

Hypothesis 2. *For the number of bent functions in n variables,*

$$2^{2^{n-2}+\frac{1}{4}\binom{n}{n/2}} \leqslant |\mathcal{B}_n| \leqslant 2^{2^{n-1}+\frac{1}{2}\binom{n}{n/2}}.$$

Thus, we suppose that the number of all bent functions is very close to the existing upper bound.

Hypothesis 3. *The number of all bent functions in n variables (n is even, $n \geqslant 2$) is asymptotically equal to $2^{2^{n-c}+d\binom{n}{n/2}}$, where c and d are constants and $1 \leqslant c \leqslant 2$.*

If Hypothesis 1 is correct, then using Theorem 91, one can prove the following hypothesis:

Hypothesis 4. *The class $\mathcal{BI}_n$ is the basic class in $\mathcal{B}_n$—that is,*

$$\lim_{n\to\infty} \frac{\log\log|\mathcal{BI}_n|}{\log\log|\mathcal{B}_n|} = 1.$$

One can see these relations for small n in Table 13.3.

If n is small, the dynamics of the corresponding relations is not still impressive. But it can be explained by an *effect of small values.*

We saw in Section 13.7 that the lower bound for $|\mathcal{BI}_n|$ from Theorem 91 is very close to the exact value if n is small. We make the following assumption:

Hypothesis 5. *The bound of Theorem 91 is asymptotically exact—that is,*

$$\lim_{n\to\infty} \frac{\log\log(|\mathcal{B}_{n-2}|^4/|X_{n-2}|)}{\log\log|\mathcal{BI}_n|} = 1.$$

One can see the values $\log\log(|\mathcal{B}_{n-2}|^4/|X_{n-2}|)$ and $\log\log|\mathcal{BI}_n|$ for small n in Table 13.4. They are indeed very close to each other.

The results obtained lead us to a new vision of the enumeration problem for bent functions. Hypotheses 1, 4, and 5 give the following directions for further study of the problem. First, to study the size of X_n. If it is big enough, it is possible to get a good lower bound for the number of bent functions. Second, to study the distributions of the multiplicities in X_n in order to

Table 13.3 Relations between $\log\log|\mathcal{BI}_n|$ and $\log\log|\mathcal{B}_n|$ if n is small

	$n = 2$	$n = 4$	$n = 6$	$n = 8$
$a = \log\log|\mathcal{BI}_n|$	≈ 1.584962501	≈ 3.169925001	≈ 4.821035977	≈ 6.449066085
$b = \log\log|\mathcal{B}_n|$	≈ 1.584962501	≈ 3.293864089	≈ 5.015117973	≈ 6.731862061
a/b	1	0.962372738	0.961300612	0.957991419

find the number of bent iterative functions. This number by Hypothesis 5 is asymptotically equal to the number of all bent functions. It is interesting also to study several weakened variants of the hypotheses.

Table 13.4 Relations between $\log\log(|\mathcal{B}_{n-2}|^4/|X_{n-2}|)$ and $\log\log|\mathcal{BI}_n|$ if n is small

	$n = 4$	$n = 6$	$n = 8$
$a = \log\log(|\mathcal{B}_{n-2}|^4/|X_{n-2}|)$	≈ 3.169925001	≈ 4.819127567	≈ 6.448708743
$b = \log\log|\mathcal{BI}_n|$	≈ 3.169925001	≈ 4.822730148	≈ 6.449066085
a/b	1	0.999604149	0.99994459

CHAPTER 14

Bent Decomposition Problem

INTRODUCTION

The following problem is considered in this chapter: Is it true that an arbitrary Boolean function in n variables (n is even, $n \geqslant 2$) of degree not more than $n/2$ can be represented as the sum of two bent functions in n variables? This question was raised in 2011 in close connection with the problem of the asymptotic value of the number of all bent functions in n variables. There is a hypothesis that the answer is "yes." For now all known facts confirm the hypothesis. The hypothesis was checked for $n = 2, 4, 6$ and for some special classes of Boolean functions. Some weakened variants of it are also proven. In this chapter we discuss what is known in this area.

14.1 PRELIMINARIES

One of the most important problems in bent functions is to find the number of them. In Chapter 13 (see Sections 13.6 and 13.8) a new approach to this problem was discussed and the following hypothesis was formulated:

Hypothesis 1. *Every Boolean function in n variables of degree not more than n/2 can be represented as the sum of two bent functions in n variables (n is even, $n \geqslant 2$).*

Firstly, this hypothesis was stated by the author in 2011 [363]. This hypothesis is an analogue of Goldbach's conjecture in number theory, which is unsolved since 1742: any even number $n > 4$ can be represented as the sum of two prime numbers. If one can prove the stated hypothesis on bent functions, then the asymptotic value of the number of all bent functions will be found, and hence an answer to the main question in bent functions will be given.

In [363] we checked the hypothesis for Boolean functions in n variables for all possible small cases: for $n = 2, 4, 6$. To check the case $n = 8$ is too hard, since there is no complete affine classification of Boolean functions of degree 4 in eight variables. It is shown in [365] that any cubic Boolean function in eight variables is the sum of not more than four bent functions in eight variables.

Bent Functions
http://dx.doi.org/10.1016/B978-0-12-802318-1.00014-5

In 2014 Qu et al. [312] confirmed the hypothesis in some particular cases. Namely, they proved that all quadratic Boolean functions, Maiorana-McFarland bent functions, and partial spread functions can be represented as the sums of two bent functions. In [365] a weakened variant of Hypothesis 1 was studied by the author. It was proven that every Boolean function in n variables of a constant degree d, where $d \leqslant n/2$, can be represented as the sum of a constant number of bent functions in n variables.

14.2 PARTIAL RESULTS

In Chapter 13 we saw that for $n = 2, 4, 6$ the set X_n contains all Boolean functions of degree not more than $n/2$. Recall that $X_n = \{ f \oplus h : f, h \in \mathcal{B}_n \}$. The case $n = 6$ was checked via an exhaustive search. We used an Intel Core i7 3.0 GHz processor 256 GB of RAM. The program worked 14 days with full loading of RAM. Case $n = 8$ is too hard for an exhaustive search. In this case one has to find bent decompositions for about 2^{163} Boolean functions in eight variables. Recall that the number of all bent functions in eight variables is about $2^{106.29}$. Case $n = 10$ cannot be checked even theoretically since the number of bent functions in ten variables is unknown. Thus, we have the following theorem [363]:

Theorem 96. *Hypothesis 1 is true for $n = 2, 4, 6$.*

Recall that a Boolean function f in n variables *depends on variable x_i* if its algebraic normal form contains x_i. Qu et al. [312] proved the following theorem:

Theorem 97. *The following statements hold:*

(1) *Every Maiorana-McFarland bent function can be represented as the sum of two bent functions.*

(2) *Every Boolean function in n variables that depends on $n/2$ variables (or fewer) is the sum of two bent functions.*

(3) *Every quadratic Boolean function is the sum of two bent functions.*

Proof.

(1) Indeed, if one considers a Boolean function in n variables as a function from $\mathbb{F}_{2^n}$ to $\mathbb{F}_2$, then $f(x, y) = \langle x, \pi(y) \rangle \oplus h(y)$ can be represented as the sum of two bent functions f_1 and f_2, where $f_1(x, y) = \langle x, \beta\pi(y) \rangle$ and $f_2(x, y) = \langle x, (\beta + 1)\pi(y) \rangle \oplus h(y)$, where β is an arbitrary element from $\mathbb{F}_{2^n} \setminus \{0, 1\}$. Note that in the McFarland construction, n variables can be partitioned into halves in an arbitrary way.

(2) If a function f depends only on variables $x_1, \ldots, x_d$, $d \leqslant n/2$, then it is the sum of two McFarland bent functions $g(x, y) = \langle x, y \rangle$ and

$h(x, y) = \langle x, y \rangle \oplus f'(x)$, where $f'(x) = f(x, y)$ and variables $x_1, \ldots, x_d$ are covered by vector x.

(3) The proof of this fact is based on the known affine classification of all quadratic Boolean functions in n variables (because of Dickson's theorem) and can be found in [312].

□

Qu and Li [312] also proved that partial spread Boolean functions can be decomposed into the sum of two bent functions.

14.3 BOOLEAN FUNCTION AS THE SUM OF A CONSTANT NUMBER OF BENT FUNCTIONS

In this section we prove a weakened variant of Hypothesis 1 following [365].

Theorem 98. *Every Boolean function in n variables of degree d, where $3 \leqslant d \leqslant n/2$, n is even, can be represented as the sum of a constant number N_d of bent functions in n variables. Moreover, $N_d \leqslant 2\binom{2k}{d}$, where k is the least number, $k \geqslant d$, such that $n/2$ can be divided by k.*

Proof. Let f be a Boolean function of degree d. Consider two cases.

Case 1. Let d divide $n/2$—that is, $n = 2dm$ for some integer m. Then the number k defined in the statement of the theorem is equal to d. Consider the partition of the set $\{1, 2, 3, \ldots, n\}$ into $2d$ subsets:

$$A_1 = \{1, \ldots, m\}, \; A_2 = \{m + 1, \ldots, 2m\},$$

$$A_3 = \{2m + 1, \ldots, 3m\}, \; \ldots, \; A_{2d} = \{2m(d - 1) + 1, \ldots, n\}.$$

For any monomial of degree up to d it is possible to choose d subsets A_i in such a way that union of them—say, A—covers all variables of the monomial. Note that $|A| = n/2$. Every such choice of d subsets A_i among the given $2d$ sets defines a partition of all variables into halves. Let x' be the vector of variables with numbers from A and let x'' be the vector of variables with numbers from $\{1, 2, 3, \ldots, n\} \setminus A$. There are $\binom{2d}{d}$ such partitions (x', x'') of n variables. Here partitions (x', x'') and (x'', x') are distinct. We say that a monomial is *associated* with a partition (x', x'') if all its variables are covered by x'. Divide all monomials of the Boolean function f into groups by the association with the same partition (x', x''). One can see that the sum of monomials of one group is a Boolean function that depends on not more than $n/2$ variables. Then by Theorem 97, item (2), it can be represented as the sum of two bent functions in n variables.

Thus, every Boolean function f can be expressed as the sum of not more than $2\binom{2d}{d}$ bent functions.

Case 2. The number d does not divide $n/2$. Take the least k such that $k > d$ and $n = 2k\ell$, where $1 \leqslant \ell \leqslant n/8$. Operations similar to those in case 1 can be performed. Let us construct partitions (x', x'') using sets:

$$B_1 = \{1, \ldots, \ell\}, \quad B_2 = \{\ell + 1, \ldots, 2\ell\},$$

$$B_3 = \{2\ell + 1, \ldots, 3\ell\}, \quad \ldots, \quad B_{2k} = \{2\ell(k-1) + 1, \ldots n\}.$$

Namely, for every monomial of degree not more than d one can choose d subsets B_i in such a way that union of them (denote it by B) covers all variables of the monomial. Note that $|B| = d\ell < n/2$. If we add to the set B any other $k - d$ subsets B_i (say, with the smallest possible indices), then we obtain the set of size $n/2$ (it is considered to be the set of coordinates of x') and hence get a partition (x', x'') of all variables into halves. Thus, using not more than $\binom{2k}{d}$ such partitions (x', x''), we can represent a Boolean function f as the sum of at most $2\binom{2k}{d}$ bent functions in the same way as in case 1.

□

By Theorem 98 it is easy to derive the following concrete facts:
- Let even n be a multiple of 3. Then every cubic Boolean function in n variables is the sum of not more than 40 bent functions in n variables.
- Let $n/2$ be a multiple of 4. Then an arbitrary Boolean function in n variables of degree 4 is the sum of not more than 140 bent functions in n variables.
- Let even n be a multiple of 5. Then every Boolean function in n variables of degree 5 is the sum of not more than 504 bent functions in n variables.

The next very simple statement is devoted to decompositions of bent functions. It shows the speciality of the case $k = 2$ in decomposition of a bent function into the sum of k bent functions [365].

Theorem 99. *Let k be a positive integer, $k \neq 2$. An arbitrary bent function in n variables can be represented as the sum of k distinct bent functions in n variables.*

Proof. Let f be a bent function in n variables. Let us show how to get the required decompositions. Case $k = 1$ is not of interest to us. Consider

$k = 3$. Take an arbitrary bent function g in n variables. Let g_ℓ be a bent function obtained from g by adding an affine (nonzero) Boolean function ℓ. Then $f = f_\ell \oplus g \oplus g_\ell$. It is easy to see that by considering another bent function h and its shift $h_{\ell'}$ (that cannot be obtained from f and g by adding an affine function), one can get a decomposition of f into five distinct bent functions, $f = f_{\ell \oplus \ell'} \oplus g \oplus g_\ell \oplus h \oplus h_{\ell'}$, and so on.

If k is even, consider a bent function g from the McFarland class. By Theorem 97, item (2), there are bent functions m and m' such that $g = m \oplus m'$. Then we get a bent decomposition for the function f into four bent functions—namely, $f = f_\ell \oplus m \oplus m' \oplus g_\ell$. It is clear now how to get decompositions of f into six or more bent functions. $\quad\square$

14.4 ANY CUBIC BOOLEAN FUNCTION IN EIGHT VARIABLES IS THE SUM OF AT MOST FOUR BENT FUNCTIONS

In this section we find bent decompositions of Boolean functions in eight variables of degree up to 3.

Recall that Boolean functions f and g in n variables are *extended affinely equivalent* if there is a nonsingular binary $n \times n$ matrix A, vectors u and v of length n, and constant $\lambda \in \mathbb{Z}_2$, such that $g(x) = f(Ax \oplus u) \oplus \langle v, x \rangle \oplus \lambda$. We can study bent decompositions only for nonequivalent Boolean functions for the following reasons:

- A Boolean function extended affinely equivalent to a bent function is bent too (Theorem 17).
- If a Boolean function f in n variables can be represented as the sum of k bent functions, then every Boolean function equivalent to f can also be represented as the sum of k bent functions.

We use the known extended affine classification of Boolean functions in eight variables up to degree 3; it can be found, for example, in [87, 243, 246]. In 2014 the following theorem was proven [365]:

Theorem 100. *Every cubic Boolean function in eight variables is the sum of not more than four bent functions.*

Proof. Consider all extended affinely nonequivalent bent functions in eight variables of degree not more than 3. We list them in the following table according to the classification given in Section 7.3. For brevity, we write monomial $x_1 x_2 x_3$ as 123 and so on. Let $f(x) = f_3(x) \oplus f_2(x)$ be an arbitrary cubic Boolean function in eight variables, where $f_3(x)$ is a homogeneous part of degree 3 and $f_2(x)$ has degree 2 or less. Without loss of generality, assume that f_3 is from the table on pages 129–130 (otherwise consider a function affinely equivalent to f).

It is not hard to get decompositions of the Boolean function f up to the quadratic part. It is enough to use only following nonequivalent bent functions:

$a = 123 + 14 + 25 + 36 + 78;$
$b = 123 + 145 + 34 + 16 + 27 + 58;$
$c = 123 + 145 + 346 + 35 + 16 + 15 + 27 + 48;$
$d = 123 + 347 + 356 + 14 + 76 + 25 + 45 + 38;$
$e = 123 + 145 + 247 + 346 + 35 + 17 + 25 + 26 + 48.$

We give the required decomposition of nonequivalent Boolean functions in eight variables up to degree 3 in the form $f(x) = g(\pi(x)) \oplus h(\sigma(x)) \oplus q(x)$, where g and h are bent functions from the set $\{a, b, c, d, e\}$, substitutions π and σ are nonsingular affine transformations of variables (permutations in most cases), and function q is a certain Boolean function of degree 2 or less (we do not concretize it). According to Theorem 97, item (3), every quadratic function q is the sum of two bent functions. Thus, f can be represented as the sum of not more than four bent functions in eight variables.

For example, function $f(x) = x_1x_2x_3 \oplus x_2x_4x_6 \oplus x_3x_5x_7 \oplus x_1x_2x_8 \oplus x_1x_3x_8$ (number 15 in the following table) is the sum $b(x_2 + x_3, x_1, x_8, x_4, x_6, x_3, x_5, x_7) \oplus d(x_1 + x_2, x_2, x_3, x_4, x_5, x_7, x_6, x_8) \oplus q(x)$, where q is a quadratic function.

All the decompositions are listed in the table on pages 129–130. ☐

In fact, we suppose that for every function considered it is possible to find decompositions into exactly *two* bent functions. But it requires a more complicated technique of working with quadratic parts.

14.5 DECOMPOSITION OF DUAL BENT FUNCTIONS

Recall that according to Section 1.2 the algebraic normal form of a Boolean function f has the form

$$f(x) = \bigoplus_{y \in \mathbb{F}_2^n} f_y x_1^{y_1} \cdots \cdot x_n^{y_n}, \quad \text{where } f_y = \bigoplus_{z \in \mathbb{F}_2^n, z \preccurlyeq y} f(z).$$

The following theorem can be easily proven (see [366]):

N	Nonequivalent homogeneous Boolean functions of degree 3	g	h	π	σ
1	123	a	b	[1, 4, 5, 2, 3, 6, 7, 8]	id
2	123 + 145	a	a	id	[1, 4, 5, 2, 3, 6, 7, 8]
3	123 + 456	a	a	id	[4, 5, 6, 1, 2, 3, 7, 8]
4	123 + 135 + 236	a	b	id	[3, 1, 5, 2, 6, 4, 7, 8]
5	123 + 124 + 135 + 236 + 456	c	c	[1 + 6, 2, 3, 4, 5, 6, 7, 8]	[3 + 4, 5, 1, 4, 6, 2, 7, 8]
6	123 + 145 + 167	a	b	id	[1, 4, 5, 6, 7, 2, 3, 8]
7	123 + 246 + 357	b	d	[4, 2, 6, 3, 8, 1, 7, 5]	[1, 2, 3, 4, 5, 7, 8, 6]
8	123 + 145 + 167 + 246	a	c	id	[1, 5, 4, 6, 7, 2, 3, 8]
9	123 + 145 + 246 + 357	d	d	[1, 2, 3, 4, 5, 7, 8, 6]	[1, 5, 4, 2, 3, 8, 6, 7]
10	123 + 124 + 135 + 236 + 456 + 167	b	d	[1 + 6, 2, 3, 4, 5, 6, 7, 8]	[2 + 5, 4, 1, 3, 6, 7, 5, 8]
11	123 + 145 + 167 + 246 + 357	b	c	[6, 1, 7, 2, 4, 3, 5, 8]	[1, 2, 3, 5, 4, 7, 6, 8]
12	123 + 478 + 568	a	b	id	[8, 4, 7, 5, 6, 1, 2, 3]
13	123 + 145 + 167 + 568	a	c	id	[1, 4, 5, 6, 7, 8, 2, 3]
14	123 + 246 + 357 + 568	c	d	[4,2,6,8,3,5,1,7]	[1,2,3,4,5,7,8,6]
15	123 + 246 + 357 + 128 + 138	b	d	[2 + 3, 1, 8, 4, 6, 3, 5, 7]	[1 + 2, 2, 3, 4, 5, 7, 6, 8]
16	123 + 145 + 167 + 357 + 568	a	e	id	[1, 6, 7, 5, 4, 3, 8, 2]
17	123 + 145 + 478 + 568	a	c	id	[4, 1, 5, 8, 7, 6, 2, 3]
18	123 + 124 + 135 + 236 + 456 + 167 + 258	e	e	[1, 2 + 5, 3, 5, 4, 6, 8, 7]	[1, 2 + 5, 4, 6, 7, 5, 3, 8]
19	123 + 124 + 135 + 236 + 456 + 178	b	d	[1 + 6, 2, 3, 4, 5, 6, 7, 8]	[2 + 5, 4, 1, 3, 7, 8, 5, 6]

Continued

N	Nonequivalent homogeneous Boolean functions of degree 3	g	h	π	σ
20	123 + 145 + 246 + 357 + 568	d	e	[1, 2, 3, 4, 5, 7, 8, 6]	[5, 6, 8, 4, 1, 3, 2, 7]
21	123 + 145 + 246 + 467 + 578	c	e	[4, 3, 8, 7, 6, 5, 1, 2]	[1, 2, 3, 4, 5, 8, 6, 7]
22	123 + 145 + 357 + 478 + 568	a	e	id	[4, 7, 8, 5, 1, 6, 3, 2]
23	123 + 246 + 357 + 478 + 568	c	e	[1, 2, 3, 5, 4, 7, 6, 8]	[5, 6, 8, 4, 1, 7, 2, 3]
24	123 + 246 + 357 + 148 + 178 + 258	c	c	[1, 2, 3, 7, 8, 5, 4, 6]	[2, 5, 8, 4, 6, 1, 3, 7]
25	123 + 145 + 167 + 246 + 357 + 568	c	d	[1, 2, 3, 5, 4, 7, 6, 8]	[1, 7, 6, 2, 5, 8, 4, 3]
26	123 + 145 + 167 + 246 + 238 + 258 + 348	c	e	[1, 7 + 8, 6, 4, 5, 2, 3, 8]	[2, 1 + 8, 3, 8, 5, 4, 6, 7]
27	123 + 145 + 167 + 258 + 268 + 378 + 468	c	e	[1, 3 + 8, 2, 5, 4, 8, 6, 7]	[6, 1 + 6, 7, 8, 4, 3, 2, 5]
28	123 + 145 + 246 + 357 + 238 + 678	c	c	[1, 2, 3, 5, 4, 7, 6, 8]	[2, 3, 8, 6, 4, 7, 1, 5]
29	123 + 145 + 246 + 357 + 478 + 568	c	c	[1, 2, 3, 5, 4, 7, 6, 8]	[4, 2, 6, 8, 7, 5, 1, 3]
30	123 + 124 + 135 + 236 + 456 + 167 + 258 + 378	c	e	[1, 2, 3 + 4, 6, 7, 5, 4, 8]	[5, 8, 2 + 5, 3, 1, 6, 7, 4]
31	123 + 156 + 246 + 256 + 147 + 157 + 357 + 348 + 258 + 458	c	e	[5, 2 + 4, 8, 3, 7, 4, 1, 6]	[2, 4 + 5, 6, 1, 3, 5, 7, 8]

Theorem 101. *A bent function in n variables, $n \geqslant 4$, is decomposable into the sum of two bent functions in n variables if and only if the dual bent function is decomposable.*

Proof. Let g be a bent function in n variables such that $g = f \oplus h$, where f and h are bent functions. Then for every nonzero coefficient g_y of the algebraic normal form of the function g, $g_y = f_y \oplus h_y$, where y is an arbitrary vector. We can consider only vectors of Hamming weight less than or equal to $n/2$ as far as by Theorem 16 all coefficients g_y, f_y, h_y are equal to zero if $wt(y) > n/2$. Then we get

$$g_y = \bigoplus_{x \preccurlyeq y} g(x) = \left(\bigoplus_{x \preccurlyeq y} f(x) \right) \oplus \left(\bigoplus_{x \preccurlyeq y} h(x) \right).$$

Using the equality $a \oplus b = a + b - 2ab$, we can switch to the usual operations of addition and substraction on the right side of the equality. By Theorem 21 (see Chapter 5),

$$\sum_{x \preccurlyeq y} g(x) = 2^{wt(y)-1} - 2^{(n/2)-1} + 2^{wt(y)-n/2} \sum_{x \preccurlyeq y \oplus 1} \tilde{g}(x).$$

Using this and some other analogous equalities for functions f and h, we obtain

$$2^{wt(y)-n/2} \left(\left(\sum_{x \preccurlyeq y} \tilde{g}(x) \right) - \left(\sum_{x \preccurlyeq y} \tilde{f}(x) \right) - \left(\sum_{x \preccurlyeq y} \tilde{h}(x) \right) \right)$$

$$= 2^{wt(y)-1} - 2^{(n/2)-1} - 2f_y h_y.$$

We multiply the equality by $2^{(n/2)-wt(y)}$. Then

$$\left(\sum_{x \preccurlyeq y} \tilde{g}(x) \right) - \left(\sum_{x \preccurlyeq y} \tilde{f}(x) \right) - \left(\sum_{x \preccurlyeq y} \tilde{h}(x) \right) = 2^{(n/2)-1}$$

$$-2^{n-wt(y)-1} - 2^{(n/2)-wt(y)+1} f_y h_y.$$

Note that the expression on the right side is even since $wt(y) \leqslant n/2$ and $n \geqslant 4$. Hence, taking this equality modulo 2, we get

$$\bigoplus_{x \preccurlyeq y} \tilde{g}(x) = \left(\bigoplus_{x \preccurlyeq y} \tilde{f}(x) \right) \oplus \left(\bigoplus_{x \preccurlyeq y} \tilde{h}(x) \right);$$

that is, $\tilde{g}_y = \tilde{f}_y \oplus \tilde{h}_y$ for an arbitrary vector y of weight $\leqslant n/2$. Recall that for vectors of weight more than $n/2$ this expression automatically holds. Thus, $\tilde{g} = \tilde{f} \oplus \tilde{h}$. In the opposite direction (from $\tilde{g}$ to g), the statement can be proven in the same way.   □

There are some simple corollaries:

- Let g, f, and h be bent functions in n variables, $n \geqslant 4$. Then if $g \oplus f \oplus h = 0$, it is true $\tilde{g} \oplus \tilde{f} \oplus \tilde{h} = 0$. This means that if we know a bent decomposition of a bent function, then we easily get such a decomposition of the dual function.
- The number of distinct bent decompositions for a bent function g coincides with that number for the dual bent function $\tilde{g}$.

Snow-covered trees

CHAPTER 15

Algebraic Generalizations of Bent Functions

INTRODUCTION

Generalizations of bent functions with respect to their algebraic, combinatorial, and cryptographic properties are becoming more numerous and more widely studied from year to year. It is quite difficult not only to determine connections between generalizations, but also to collect information about all of them and briefly review the progress in this area. In this chapter and the next two chapters we provide a systematic survey of the existing generalizations of bent functions and try whenever possible to establish relations between various generalizations. In this chapter the following algebraic generalizations are considered: q-valued bent functions, p-ary bent functions, bent functions over a finite field, generalized Boolean bent functions of Schmidt, bent functions from a finite Abelian group into the set of complex numbers on the unit circle, bent functions from a finite Abelian group into a finite Abelian group, non–Abelian bent functions, vectorial G-bent functions, and multidimensional bent functions on a finite Abelian group.

15.1 PRELIMINARIES

The term *generalized bent function* is used quite often, but almost every time it means something new. Bent functions are actively studied for their numerous applications in information theory, cryptography, coding theory, and other fields. New statements of problems lead to many generalizations of bent functions, and it becomes more and more difficult to study and sort them. A systematic survey of the existing generalizations of bent functions is given in this chapter and the next two chapters. Here we follow the paper [361] published in 2011. Note that in 2015 a new survey on generalized bent functions was published by Hodžić and Pasalic [165].

We divide the generalizations of bent functions into several groups. Sometimes the division is very relative, but it seems convenient for the presentation. While describing each generalization, we pay attention if possible to

Bent Functions
http://dx.doi.org/10.1016/B978-0-12-802318-1.00015-7

- who introduced the generalization, when, and why;
- what the form of the function is and the Walsh-Hadamard (or Fourier) transform occurring as a rule in each case;
- what results are known about it;
- how this generalization is related to other generalizations, and so forth.

For every generalization we include appropriate references.

In this chapter we collect generalizations about which functions are not Boolean. As a rule, these are the mappings from one algebraic system to another.

15.2 THE q-VALUED BENT FUNCTIONS

In 1985 Kumar et al. [218] proposed this natural generalization of bent functions, aiming to construct q-valued bent sequences applicable in code division multiple access (CDMA) systems.

Take integer $q \geqslant 2$, the imaginary unit $i = \sqrt{-1}$, and a primitive complex root of unity $\omega = e^{2\pi i / q}$ of degree q. Consider a q-valued function $f : \mathbb{F}_q^n \to \mathbb{F}_q$. The *Walsh-Hadamard transform* of a function f is the complex function

$$W_f(y) = \sum_{x \in \mathbb{F}_q^n} \omega^{\langle x, y \rangle + f(x)} \quad \text{for every} \quad y \in \mathbb{F}_q^n, \tag{15.1}$$

where the inner product and addition $+$ are taken modulo q.

We denote the absolute value of a complex number c by $|c|$.

Definition 4 (Kumar et al. [218]). Given positive integer q, a function $f : \mathbb{F}_q^n \to \mathbb{F}_q$ is called a *q-valued bent function* if $|W_f(y)| = q^{n/2}$ for every $y \in \mathbb{F}_q^n$. If $q = p$, where p is a prime number, such a function is usually called a *p-ary bent function*.

Sometimes these functions are called *multiple-valued bent functions* or simply *generalized bent functions*. For $q = 2$, this concept coincides with the concept of a Boolean bent function. If $q = p$ is prime, such functions are usually called *p-ary bent functions*: we give many references to them in the next section. Here we consider the case of an arbitrary integer $q \geqslant 2$.

We denote the set of all q-valued bent functions in n variables by $\mathcal{B}_{n,q}$. As for the binary case (Theorem 17) the following statement is obtained in [218]:

Theorem 102. *The class $\mathcal{B}_{n,q}$ is closed under*

(1) *every nonsingular affine transformation of the variables;*

(2) *addition of an arbitrary q-valued affine function.*

A square $n \times n$ matrix A consisting of integer powers of ω is called a *generalized Hadamard matrix* whenever $A\overline{A}^{\mathrm{T}} = nE$, where E is the identity matrix.

Theorem 103. *The following are equivalent:*

(1) *A q-valued function f is a bent function.*

(2) $A = (a_{x,y})$ *with* $a_{x,y} = \omega^{f(x+y)}$ *is a generalized Hadamard matrix.*

Note that for $q = 2$, Theorems 102 and 103 amount to well-known facts about Boolean bent functions (see Chapter 5). The specific features of the q-valued case include the fact [218] that f remains a bent function when we replace ω in the definition of $W_f(y)$ by another primitive root of unity γ of degree q. Note also that q-valued bent functions exist for both even and odd n. Let us consider several constructions.

Theorem 104. *Take arbitrary positive integers m, n, and q. For arbitrary functions $g \in \mathcal{B}_{m,q}$ and $h \in \mathcal{B}_{n,q}$, the function $f(x', x'') = g(x') + h(x'')$ is a q-valued bent function in $m + n$ variables.*

An analogue of the Maiorana-McFarland construction holds:

Theorem 105. *If n is even and q is arbitrary, then*

$$f(x, y) = \langle x, \pi(y) \rangle + h(y)$$

is a q-valued bent function, where h is an arbitrary q-valued function in $n/2$ variables, and π is an arbitrary permutation on the set $\mathbb{F}_q^{n/2}$.

Suppose that n is odd, $q \equiv 2 \mod 4$ and $q > 2$. It is shown in [218] that if there is an integer b such that $2^b + 1$ is divisible by $q/2$, then there is no q-valued bent function in n variables.

Bent functions exist for every q with $q \neq 2 \mod 4$ and every n. They can be constructed using Theorem 104, for instance, from the following one-dimensional functions ($n = 1$)—see [218] for more details:

Theorem 106. *The following q-valued functions of one variable are bent functions:*

(1) $f(x) = x^2 + cx$, *where $c \in \mathbb{F}_q$ is an arbitrary constant (if q is odd).*

(2) $f(x) = rx'h(x'') + g(x'')$, *where $x = rx' + x'' \in \mathbb{F}_q$, $0 \leqslant x', x'' \leqslant r - 1$, h is an arbitrary permutation on $\mathbb{F}_r$, and g is an arbitrary function of the form $\mathbb{F}_r \to \mathbb{F}_q$ (if $q = r^2$ for some r).*

Several p-ary and q-valued versions of certain results regarding bent functions and resilient functions can be found in a paper by Hou [171].

The problem of the existence/nonexistence of q-valued bent functions was discussed by Akyildiz et al. [12] in 1996, and later by Feng [117], Liu et al. [242], and Feng and Liu [118], and by Liu and Yue [240, 241] in connection with diophantine equations. Note that these last authors call q-valued bent functions "generalized bent functions": we will see in Section 15.6 that this term should be used for other objects.

In 1989 Chung and Kumar [88] continued the study of q-valued bent functions. In 1998 Hou [168] proposed some constructions of q-valued bent functions via chain rings. More constructions of q-valued bent functions can be found in the papers of Carlet and Dubuc [57], Jadda et al. [189] (quaternary bent functions, 2013), and Singh et al. [329].

Regular q-valued bent functions retain the properties of Boolean bent functions completely. A bent function $f : \mathbb{F}_q^n \to \mathbb{F}_q$ is called *regular* if each of its Walsh-Hadamard coefficients can be expressed as

$$W_f(\gamma) = q^{n/2}\omega^{g(\gamma)}$$

for some q-valued function g. It can be shown [218] that g is also a regular bent function, and is called the *dual* to f.

Let us give several examples:

- *For $n = 1$ and $q = 4$, $f(x) = x^3 + 3x^2$ is a regular bent function. Its Walsh-Hadamard spectrum (the tuple of coefficients in increasing order of arguments) is*

$$(2, 2i, 2, -2i) = (2\omega^0, 2\omega^1, 2\omega^0, 2\omega^3),$$

where $\omega = e^{\pi i /2}$; the dual function $g(x)$ is equal to x^3.

- *For $n = 1$ and $q = 3$, the bent function $f(x) = x^2$ is not regular; its spectrum is equal to*

$$\left\{ \sqrt{3}i, \frac{3}{2} - \frac{\sqrt{3}}{2}i, \frac{3}{2} - \frac{\sqrt{3}}{2}i \right\}$$

or, with use of the powers of a primitive root of unity,

$$\left\{ \sqrt{3}\omega^{3/4}, \sqrt{3}\omega^{11/4}, \sqrt{3}\omega^{11/4} \right\},$$

where $\omega = e^{2\pi i/3}$. Here all exponents in the powers of ω are fractional.

It is not difficult to see that a Boolean bent function ($q = 2$) is always regular. The bent functions constructed in Theorems 105 and 106 (for $q \equiv 1$ mod 4 in item (1)) are regular. For odd n and $q \equiv 2, 3$ mod 4, no regular bent function exists [218]. Agievich [9] showed that regular q-valued bent functions can be described using *bent rectangles*; for the binary case, see the description in Section 6.7.

In 2009 q-valued bent functions for $q = 4$ were studied by Solé and Tokareva [331].

In 2012 new classes of generalized bent functions over $\mathbb{Z}_4$ were constructed by Li et al. [235].

Bent functions over rings were studied by Teng et al. [352, 353]. Carlet et al. [64] have proposed a class of bent functions on a Galois ring. On the basis of this class, systematic authentication codes were presented [64].

In 2012 Budaghyan et al. [27] analyzed most of the known infinite classes of q-valued bent functions for their relation to the completed Maiorana-McFarland class. This was done using the criterion based on second-order derivatives of a function. They showed that, unlike in the binary case, not all quadratic q-valued bent functions are extended affinely equivalent to a function of the Maiorana-McFarland type. They mention this result as the first attempt to state the problem for the generalized bent functions.

15.3 THE p-ARY BENT FUNCTIONS

Here we collect additional information about q-valued bent functions in the case when $q = p$ is prime. Such functions are usually called p-ary bent functions.

In 2004 Hou [171] proved the following important theorem:

Theorem 107. *Let p be a prime number. If f is a p-ary bent function in n variables, then*

$$\deg(f) \leqslant \frac{(p-1)n}{2} + 1.$$

In addition, if f is weakly regular, then

$$\deg(f) \leqslant \frac{(p-1)n}{2}.$$

In 2012 the question of attaining these bounds was discussed by Çeşmelioğlu and Meidl [72].

Constructions of p-ary bent functions are discussed in the papers of Kim et al. [201], Li et al. [237] (quadratic p-ary bent functions), Zheng et al. [401], and Çeşmelioğlu et al. [71] (constructions of weakly and non-weakly regular p-ary bent functions from semibent functions). The generalized Maiorana-McFarland class and normality of p-ary bent functions were discussed by Çeşmelioğlu et al. [74].

Several conjectures on ternary weakly regular bent functions and their proofs were considered by Helleseth et al. [156].

Duals of monomial quadratic p-ary bent functions were studied by Helleseth and Kholosha [158]. Functions dual to a Coulter-Matthews bent function (ternary bent function of the special type) were studied in 2007 by Hou [172]. Functions dual to certain ternary weakly regular bent functions were discussed by Gong et al. [144] in 2012. In 2013 Çeşmelioğlu et al. [73] studied the problem of "bentness" for functions dual to p-ary bent functions.

In 2010 Tan et al. [348] proved a new characterization of weakly regular ternary bent functions via partial difference sets (that correspond to strongly regular graphs, as is known). Using known families of bent functions, they obtained new families of strongly regular graphs, some of which were previously unknown. Moreover, they gave a new proof that the Coulter-Matthews and ternary quadratic bent functions are weakly regular [348]. Generalization of this technique from the ternary case to an arbitrary p-ary case was proposed in 2011 by Chee et al. [84].

In 2010 Helleseth and Kholosha [159] studied p-ary functions of the special trace form with respect to their exponential sums. An excellent survey of cross-correlation of m-sequences, exponential sums, p-ary bent functions and Jacobsthal sums was done by them in 2011 [161]. Association schemes arising from p-ary bent functions were studied by Pott et al. [308] in 2011.

Several papers are written in terms of *multiple-valued bent functions*. Usually, this term is used for p-ary bent functions.

In 2011 Stankovic et al. [343] considered the representation of multiple-valued bent functions using Vilenkin-Chrestenson decision diagrams. The possibility of obtaining multiple output bent functions from certain power polynomials over finite fields were considered in 2012 by Pasalic and Zhang [298]. In 2013 Moraga et al. [279] also studied multiple-valued bent functions: a class of bent functions, called *strict bent functions*, was introduced, and its characterization was given. Multiple-valued hyperbent functions were studied by Moraga et al. [280].

In 2013 Hou [174] proposed a classification of self-dual quadratic p-ary bent functions when p is an odd prime number for an arbitrary n. He classified all self-dual quadratic bent functions from $\mathbb{F}_p^n$ to $\mathbb{F}_p$ under the action of the orthogonal group $O(n, \mathbb{F}_p)$. The sizes of the $O(n, \mathbb{F}_p)$-orbits of such self-dual bent functions were explicitly determined.

In 2014 Lisonek and Lu [239] considered generalization of the partial spread construction to the p-ary case. They constructed two classes of bent functions from $\mathbb{F}_p^n$ to $\mathbb{F}_p$. Their constructions generalize the classes $\mathcal{PS}^-$ and $\mathcal{PS}^+$. Also in 2014, necessary conditions for the existence of regular p-ary bent functions were discussed by Hyun et al. [181], and Bajric et al. [15] studied p-ary bent functions with Dillon's exponents.

Some symmetric factorizations of bent functions were considered by Matsufuji and Suehiro [256] in 1999.

15.4 BENT FUNCTIONS OVER A FINITE FIELD

In 1994, Ambrosimov [13] proposed another, probabilistic, definition of q-valued bent functions. In contrast to Section 15.2, here we consider only the q-valued functions over the finite field $\mathbb{F}_{q^n}$.

Suppose that $q = p^\ell$, where p is prime and ℓ is a positive integer. Take the primitive complex root of unity $\omega = e^{2\pi i/p}$ of degree p.

Take a q-valued function $f : \mathbb{F}_{q^n} \to \mathbb{F}_q$. Assume that a vector $x \in \mathbb{F}_{q^n}$ is chosen randomly and equiprobably. For the random variable $\xi = f(x)$, define the characteristic function

$$\varphi_\xi(z) = \mathbf{E}\,\omega^{\langle \xi, z \rangle}, \quad z \in \mathbb{F}_q,$$

regarding ξ and z as vectors of length ℓ over the prime field $\mathbb{F}_p$ and taking the inner product $\langle \xi, z \rangle$ modulo p. For fixed $z \in \mathbb{F}_q$, the *Walsh-Hadamard transform* of f is defined as

$$W_{f,z}(\gamma) = q^n\,\varphi_{\langle x, \gamma \rangle + f(x)}(z),$$

or, which is the same,

$$W_{f,z}(\gamma) = q^n\,\mathbf{E}\,\omega^{\langle \langle x, \gamma \rangle + f(x), z \rangle} \quad \text{for every} \quad \gamma \in \mathbb{F}_{q^n},$$

where we take the inner product $\langle x, \gamma \rangle$ modulo q. Expanding the expectation, we obtain

$$W_{f,z}(\gamma) = \sum_{x \in \mathbb{F}_{q^n}} \omega^{\langle \langle x, \gamma \rangle + f(x), z \rangle} \quad \text{for} \quad \gamma \in \mathbb{F}_{q^n}. \tag{15.2}$$

Note that in (15.1) and (15.2) we use the primitive roots of unity of different degrees q and p, respectively. The parameter z in (15.2) determines the projection of $\langle x, \gamma \rangle + f(x)$ from $\mathbb{F}_q$ to the prime field $\mathbb{F}_p$.

We can propose an equivalent definition

$$W'_{f,z}(y) = \sum_{x \in \mathbb{F}_{q^n}} \omega^{\mathrm{tr}(\langle x,y \rangle + zf(x))}$$

replacing the inner product by the trace function $\mathrm{tr} : \mathbb{F}_q \to \mathbb{F}_p$. With this definition, $W_{f,z}(y)$ and $W'_{f,z}(y)$ differ only up to a permutation on the components of z and y.

According to [13], every function f and every nonzero z satisfy Parseval's equality $\sum_{y \in \mathbb{F}_{q^n}} |W_{f,z}(y)|^2 = q^{2n}$, which implies $\max_{y \in \mathbb{F}_{q^n}} |W_{f,z}(y)| \geqslant q^{n/2}$.

Definition 5 (Ambrosimov [13]). Take $q = p^\ell$ with prime p. A function $f : \mathbb{F}_{q^n} \to \mathbb{F}_q$ is called a *bent function* if, for all $z \in \mathbb{F}_q \backslash \{0\}$ and $y \in \mathbb{F}_{q^n}$,

$$|W_{f,z}(y)| = q^{n/2}.$$

Let us make some remarks:

- For $q = p$ and $\ell = 1$, Definition 5 of q-valued bent functions coincides with Definition 4 (Kumar et al. [218]).
- In Definition 5, the Walsh-Hadamard coefficients must be equal in absolute value for every nonzero projection of the exponent of the power of the primitive element in (15.2) from $\mathbb{F}_q$ to the field $\mathbb{F}_p$. Then, as in Definition 4, they are equal in absolute value without considering the projections (moreover, $\mathbb{F}_q$ need not be a field).

Let us present several examples. Every q-valued function $f(x) = a_2 x^2 + a_1 x + a_0$ of one variable, where $a_2 \neq 0$ and $p \neq 2$, is a bent function in the sense of Ambrosimov. Every function $f(x_1, x_2) = x_1 x_2 + a_2 x_1^2 + b_2 x_2^2 + a_1 x_1 + b_1 x_2 + c$ of two variables over a field of characteristic 2 is a bent function.

For bent functions over a field, we have Rothaus's criterion [13]:

Theorem 108. *A function $f : \mathbb{F}_{q^n} \to \mathbb{F}_q$ is a bent function if and only if, for every fixed $y \in \mathbb{F}_{q^n}$, the function $f(x + y) - f(x)$ is uniformly distributed on $\mathbb{F}_q$ whenever the argument x is uniformly distributed on $\mathbb{F}_{q^n}$.*

A description of all quadratic q-valued bent functions in n variables is given in [13], where their number (we denote it by $M_q(n)$) is also calculated.

Theorem 109. *Take $q = p^\ell$. The following hold:*

(1) *If $p = 2$ and $\ell \geqslant 2$, then*

$$M_q(n) = \begin{cases} q^{\binom{n}{2}+2n+1} \prod_{j=1}^{n/2}(1 - q^{-2j+1}), & \text{for even } n, \\ 0, & \text{for odd } n. \end{cases}$$

(2) *If $p \neq 2$, then*

$$M_q(n) = (q-1)q^n M_q(n-1) + q^{n+1}(q^{n-1} - 1)M_q(n-2) \text{ for } n \geqslant 3.$$

But an explicit relationship between the Ambrosimov bent functions and those of Kumar et al. is not described in [13]. For $q = p^\ell$, it is not clear, for instance, whether a bent function in one sense is a bent function in the other sense.

Construction of bent functions over finite fields from partial spreads was proposed in 2002 by Kim et al. [199]; see also [198].

Monomial and quadratic bent functions over the finite fields of odd characteristic were studied by Helleseth and Kholosha [157] in 2005. Later, they [160] proposed new p-ary weakly regular bent functions of the form $f(c) = \mathrm{tr}_{4k}(c^{p^{3k}+p^{2k}-p^k+1} + c^2)$ mapping $\mathbb{F}_{p^{4k}}$ to $\mathbb{F}_p$. These were the first proven infinite class of nonquadratic generalized bent functions over the fields of an arbitrary odd characteristic.

In 2011 several characterizations of bent and almost bent functions on $\mathbb{F}_p^2$ were proposed by Zhang et al. [394].

In 2012 Jia et al. [190] proposed a class of binomial bent functions over the finite fields of odd characteristic. For further information on this topic, see also the paper by Zheng et al. [399].

15.5 BENT FUNCTIONS OVER QUASI-FROBENIUS LOCAL RINGS

As mentioned by Hou [169], bent functions and partial difference sets have been constructed from finite principal ideal local rings. In 2000 Hou generalized the constructions to finite quasi-Frobenius local rings. Let R be a finite quasi-Frobenius local ring with maximal ideal M. Bent functions and certain partial difference sets on $M \times M$ are then extended to $R \times R$; for the details, see [169].

15.6 GENERALIZED BOOLEAN BENT FUNCTIONS (OF SCHMIDT)

In 2006, Schmidt [323] considered another generalization of bent functions in connection with constructions of quaternary constant-amplitude codes for multicode CDMA systems. Let us consider some details of it.

Recall that CDMA is a technology for digital mobile service; for more details, see Section 4.9. In 2000, Wada [369] established a connection between bent functions and codes for CDMA.

Consider again the simplest model of information transmission in a multicode CDMA system. For a power of two $N = 2^n$, take a size $N \times N$ Hadamard matrix $A_N = (a_{jt})$ of Sylvester type. There are N parallel data flows. We can represent the transmitted information as a binary vector c of length N (one bit from each flow). The signal in multicode CDMA is modeled as

$$S_c(t) = \sum_{j=0}^{N-1} (-1)^{c_j} a_{jt},$$

where $t = 0, 1, \dots, N-1$ is a discrete time parameter—that is, the jth row of the matrix A is multiplied by $(-1)^{c_j}$, and the transmitted signal S_c is the sum of these new rows. At every moment of time, one bit of the sequence S_c is transmitted. An important parameter is the *peak-to-average power ratio* (PAPR) of the signal, which is defined as

$$\text{PAPR}(c) = \frac{1}{N} \max_t |S_c(t)|^2.$$

Note that $1 \leqslant \text{PAPR}(c) \leqslant N$. The quantity $|S_c(t)|^2$ is proportional to the power necessary to transmit this signal; thus, the vectors c with minimal $\text{PAPR}(c)$ are most suitable for transmission. We may assume that the vectors c are chosen from some binary code C of length N. Put

$$\text{PAPR}(C) = \max_{c \in C} \text{PAPR}(c).$$

If $\text{PAPR}(C) = 1$, then C is called a *constant-amplitude code*. Currently, it is a problem to construct a code of this type with large size and large code distance. Note that a code C of length 2^n is a constant-amplitude code if and only if every code word is a vector of values of some bent function in n variables [300, 369]. Indeed, given the vector c of values of a Boolean function f in n variables, $\text{PAPR}(c) = \frac{1}{2^n} \max_{x \in \mathbb{F}_2^n} |W_f(x)|^2$. Therefore, bent functions play a substantial role in constructing codes for CDMA systems.

The generalization due to Schmidt [323] is as follows: For an integer $q \geqslant 2$, take a primitive complex root of unity $\omega = e^{2\pi i / q}$ of degree q. A function $f : \mathbb{F}_2^n \to \mathbb{F}_q$ is called a *generalized Boolean function*. Its *Walsh-Hadamard transform* is the complex function

$$W_f(y) = \sum_{x \in \mathbb{F}_2^n} (-1)^{\langle x, y \rangle} \omega^{f(x)} \quad \text{for every} \quad y \in \mathbb{F}_2^n.$$

Definition 6 (Schmidt [323]). For integer q, a function $f : \mathbb{F}_2^n \to \mathbb{F}_q$ is called a *generalized bent function* (or a *gbent function*) if $|W_f(\gamma)| = 2^{n/2}$ for every $\gamma \in \mathbb{F}_2^n$.

These functions are used for constructing the constant-amplitude codes for the q-valued version of multicode CDMA, which models a binary vector c of length N as

$$S_{c,q}(t) = \sum_{j=0}^{N-1} \omega^{c_j} a_{jt}.$$

Note also that, for some problems concerning cyclic codes, Schmidt's definition seems more natural than the definition of q-valued bent functions of Kumar et al.

Schmidt [323] considers in detail the case $q = 4$, studies the relations between generalized bent functions, constant-amplitude codes, and the available $\mathbb{Z}_4$-linear codes.

An interesting question remains: How related to each other are the bent functions of Schmidt, the q-valued bent functions, and the Boolean bent functions?

This question is answered in [331] in one particular case. Suppose that a generalized Boolean function $f : \mathbb{F}_2^n \to \mathbb{F}_4$ ($q = 4$) can be represented as $f(x) = a(x) + 2b(x)$, where a and b are Boolean functions in n variables. It is shown in [331] that f is a generalized bent function if and only if b and $a + b$ are bent functions.

Note that the *real-valued bent functions* of the form $\mathbb{F}_2^n \to \{0, 1/2, 1, 3/2\}$ considered by Matsufuji and Imamura [254] coincide with the generalized bent functions for $q = 4$.

In 2013 Stănică et al. [342] investigated the properties of generalized bent functions defined on $\mathbb{F}_2^n$ with values in $\mathbb{F}_q$, where $q \geqslant 2$ is a positive integer. They characterized the class of generalized bent functions symmetric with respect to two variables, and provided analogues of Maiorana-McFarland-type bent functions and generalized Dillon's functions. For functions from $\mathbb{F}_2^n$ to $\mathbb{F}_4$ and $\mathbb{F}_8$ several constructions and characterizations were also proposed.

In 2014 Li et al. [236] proposed new constructions of generalized Boolean bent functions and p-ary bent functions.

15.7 BENT FUNCTIONS FROM A FINITE ABELIAN GROUP INTO THE SET OF COMPLEX NUMBERS ON THE UNIT CIRCLE

In 1997, Logachev et al. [244] introduced the concept of bent functions on an arbitrary finite Abelian group. In the case of an elementary Abelian 2-group, this concept coincides with the concept of Boolean bent functions.

Take a finite Abelian group $(A, +)$ of order n, the maximal order of whose elements (the *exponent* of the group) is equal to q. Denote the group of degree q roots of unity by

$$T_q = \{e^{2\pi i\, k/q} \mid k = 0, 1, \ldots, q - 1\},$$

and the group of homomorphisms $\chi : A \to T_q$, by $\hat{A}$, which is called the *character group* of A (or its *dual group*). It is known that A and $\hat{A}$ are isomorphic. Fix some isomorphism $y \in A$, $y \to \chi_y$.

Instead of the Walsh-Hadamard transform, it is convenient to introduce the *Fourier transform* of a complex valued function $f : A \to \mathbb{C}$:

$$\hat{f}(y) = \sum_{x \in A} f(x)\overline{\chi_y(x)}.$$

Henceforth, we consider only the functions from A into $\mathbb{C}$ all of whose values lie on the unit circle $S_1(\mathbb{C})$ centered at the origin.

Definition 7 (Logachev et al. [244]). Take a finite Abelian group A of order n. A function $f : A \to S_1(\mathbb{C})$ is called a *bent function* if $|\hat{f}(y)|^2 = n$ for every $y \in A$.

We make the following remarks:

- If A is an elementary Abelian 2-group—that is, $q = 2$ and $n = 2^m$ for some positive integer m—then this concept coincides with the concept of ordinary bent functions of m variables.

- Take two integers q and m. Then the q-valued bent functions of m variables of Kumar et al. (see Definition 4) constitute a particular case of the bent functions of Definition 7 with $A = \mathbb{F}_q^m$ and $n = q^m$, but a little modification is necessary: from the functions of the form $f : \mathbb{F}_q^m \to \mathbb{F}_q$ we have to pass to the functions $f' : \mathbb{F}_q^m \to T_q \subset \mathbb{C}$, where $f'(x) = \omega^{f(x)}$. As an isomorphism between A and its character group $\hat{A}$, choose the correspondence $y \to \chi_y(x) = \omega^{\langle x, y \rangle}$, where $\omega = e^{2\pi i /q}$. A function $f : A \to S_1(\mathbb{C})$ is called *balanced* whenever

$$\sum_{x \in A} f(x) = 0.$$

The following is the Rothaus criterion for bent functions on a group [244]:

Theorem 110. *A function f is a bent function on the group A if and only if* $\overline{f(x)}f(x+y)$ *is balanced for every* $y \in A$, $y \neq 0$.

Some other criteria can be found in [244].

As in the case of a Boolean function, for a bent function f on a group A, we can define the *dual* function $\tilde{f} : A \to S_1(\mathbb{C})$ by

$$\tilde{f}(x) = \frac{1}{\sqrt{n}}\hat{f}(x),$$

and $\tilde{f}$ is a bent function as well.

If for a decomposition of A into the direct product of some groups A_1 and A_2 we can express a function $f : A \to S_1(\mathbb{C})$ as

$$f(x', x'') = f_1(x')f_2(x''),$$

where $f_1 : A_1 \to S_1(\mathbb{C})$ and $f_2 : A_2 \to S_1(\mathbb{C})$, then f is called a *decomposable* function. The following theorem holds [244]:

Theorem 111. *A decomposable function f is a bent function on the group A if and only if f_1 and f_2 are bent functions on groups A_1 and A_2, respectively.*

Future ideas on this theme can be found in the paper of 2001 by Logachev et al. [245]

15.8 BENT FUNCTIONS FROM A FINITE ABELIAN GROUP INTO A FINITE ABELIAN GROUP

In 2002, Solodovnikov [332] proposed the most general approach to algebraic generalizations of bent functions. While presenting his results, we use both the original notation and that in [53], which sometimes seems more convenient. In 2004, Carlet and Ding [53] repeated the results of Solodovnikov, but unfortunately without reference to his work.

Take two finite Abelian groups $(A, +)$ and $(B, +)$ of orders n and m, respectively, with the maximal orders a and b of their elements. Let $\hat{A}$ and $\hat{B}$ denote the character groups of A and B. Fix two isomorphisms $y \to \chi_y$ and $z \to \eta_z$ between A and $\hat{A}$, as well as B and $\hat{B}$, where $\chi_y : A \to T_a$ and $\eta_z : B \to T_b$ are characters. Take an arbitrary function $f : A \to B$. We present the following definition from [332] in a slightly different form by introducing normalization factors, but this preserves its meaning.

The *Fourier transform of the character* of a function f for fixed $z \in B$ is the following function

$$\hat{f}_z(y) = \sum_{x \in A} \eta_z(f(x))\overline{\chi_y(x)}, \quad y \in A. \tag{15.3}$$

Parseval's equality $\sum_{y \in A} |\hat{f}_z(y)|^2 = n^2$ holds for every z.

Definition 8 (Solodovnikov [332]). A function $f : A \to B$ is called a *bent function* if $|\hat{f}_z(y)|^2 = n$ for all $z \in B$, $z \neq 0$, and $y \in A$.

Fixing an element $z \in B$, we can pass from f to the complex valued function $\eta_z \circ f : A \to T_b$. We can say that (15.3) is a decomposition of this function[1] with respect to the character group $\hat{A}$. Functions of the form $A \to S_1(\mathbb{C})$ have already been considered by Logachev et al. (see Definition 7). We have the following theorem [53, 332]:

Theorem 112. *A function $f : A \to B$ is a bent function if and only if $\eta_z \circ f$ for every $z \neq 0$ is a bent function in sense of Logachev et al.*

Recall that the *derivative of a function f in direction* $y \in A$ is the function $D_y f(x) = f(x + y) - f(x)$. The following theorem holds [53, 332]:

Theorem 113. *A function $f : A \to B$ is a bent function if and only if $D_y f(x)$ is a balanced function for every nonzero $y \in A$—that is, the cardinalities of all its preimages are equal.*

Suppose that f is a bent function. Then, for every linear or affine permutation π on A, the function $f \circ \pi : A \to B$ is a bent function. If $\ell : B \to C$ is a surjective linear function (where C is a finite Abelian group), then $\ell \circ f : A \to C$ is a bent function as well.

Solodovnikov [332] defined the *closeness* function of two functions $f, g : A \to B$ as

$$\delta(f,g) = \left(\frac{1}{m} \sum_{y \in B} \left(\frac{|\{x : f(x) - g(x) = y\}|}{n} - \frac{1}{m} \right)^2 \right)^{1/2}. \tag{15.4}$$

The intention is to use it to estimate the quality (or efficiency) of the replacement of one function with another. The smaller the value of $\delta(f,g)$, the less close to each other f and g are. The definition of closeness implies that $\delta(f,g) = 0$ if and only if f and g differ by a balanced function.

Let $\mathrm{Hom}(A, B)$ denote the set of all group homomorphisms from A to B. By definition, for every homomorphism h, the derivative $D_y h(x)$ in

---

[1] Here and below the expression $g \circ f(x)$ stands for the function $g(f(x))$.

every nonzero direction $\gamma \in A$ is a constant function. Then it is natural to call $f : A \to B$ such that $D_\gamma f(x)$ is balanced for every nonzero $\gamma \in A$ an *absolutely nonhomomorphic* function [332]. By Theorem 113, absolutely nonhomomorphic functions and bent functions coincide.

Theorem 114. *Given a bent function f and a homomorphism h, we have*

$$\delta(f, h) = \frac{\sqrt{m-1}}{m\sqrt{n}}.$$

In other words, a bent function is equally close to all homomorphisms. It is interesting to consider *minimal functions*, which are the least close to homomorphisms—that is, that have the minimal value of $\delta_f = \delta(f, \mathrm{Hom}(A, B))$. For $A = \mathbb{F}_q^\ell$ and $B = \mathbb{F}_q^r$, it is shown in [332] that f is a minimal function if $\delta_f = \sqrt{m-1}/(m\sqrt{n})$. A function is called *absolutely minimal* if its minimality is invariant under all epimorphisms of the group B.

Theorem 115. *For prime q, take $A = \mathbb{F}_q^\ell$ and $B = \mathbb{F}_q^r$ and suppose that bent functions from A to B exist. Then*

(1) *every bent function is absolutely minimal;*

(2) *for $q = 2$, the class of all bent functions coincides with the class of all absolutely minimal functions.*

In 2012 Solodovnikov [333, 334] continued the study of closeness. For functions from $(\mathbb{Z}/(p))^n$ to $(\mathbb{Z}/(p))^m$, where p is a prime, the property of closeness to linear functions was investigated. It was proven that, for any function, this property is inherited by its homomorphic images. As a generalization of an analogous statement for Boolean functions, it was shown that if $p = 2$ or 3, then the class of functions which are absolutely minimally close to linear ones coincides with the class of bent functions. For more on this topic, see [54].

We also mention here a paper by Chen and Polhill [86] that has a relation to bent functions and Abelian groups. Chen and Polhill presented new Abelian partial difference sets and amorphic group schemes of both Latin square type and negative Latin square type in certain Abelian p-groups. They constructed so-called *pseudoquadratic* bent functions and then used them in place of quadratic forms. A connection between strongly regular bent functions and amorphic group schemes was also discussed.

15.9 NON-ABELIAN BENT FUNCTIONS

In 2012 Poinsot [305] considered the case of non-Abelian bent functions. We say only a few words about them.

Perfect nonlinear functions from a finite group G to another group H are those functions $f : G \to H$ such that for all nonzero $\alpha \in G$ the derivative $D_\alpha f : x \to f(\alpha x) f(x)^{-1}$ is balanced. In the case where both G and H are Abelian groups, $f : G \to H$ is perfect nonlinear if and only if f is bent—that is, for all nonprincipal characters χ of H, the (discrete) Fourier transform of $\chi \circ f$ has a constant magnitude equal to $|G|$. Poinsot, using the theory of linear representations, presents similar bentness-like characterizations in the cases where G and H and G or H are (finite) non-Abelian groups. Thus, the concept of bent functions is extended to the framework of non-Abelian groups. For details, see [305].

15.10 VECTORIAL *G*-BENT FUNCTIONS

In 2002 Solodovnikov [332] suggested the idea of this generalization of the functions $f : A \to B$. In 2004, Poinsot and Harari [306] considered it in detail for the case $A = (\mathbb{Z}_2^k, +)$ and $B = (\mathbb{Z}_2^r, +)$—that is, for the vectorial Boolean functions. The generalization rests on the possibility of defining the derivative of a function $f : A \to B$ in a different fashion.

Let $S(A)$ denote the symmetric group on A in the multiplicative notation. A permutation $\sigma \in S(A)$ is called an *involution* whenever $\sigma\sigma = e$, where e is the identity permutation. A permutation σ has no *fixed points* if $\sigma(x) \neq x$ for every $x \in A$. Denote the set of all involutions σ without fixed points by $\mathrm{Inv}(A)$. A subgroup G of $S(A)$ with $G \subseteq \mathrm{Inv}(A) \cup \{e\}$ is called a *group of involutions of A.*

Suppose now that $A = \mathbb{Z}_2^k$ and $B = \mathbb{Z}_2^r$. Observe that

$$\left| \mathrm{Inv}(\mathbb{Z}_2^k) \right| = \frac{2^k!}{2^{k-1}! 2^{2^{k-1}}}.$$

It is shown in [306] that every group of involutions G of $\mathbb{Z}_2^k$ is Abelian and $|G| \leqslant 2^k$. We consider only a group G of the maximal order 2^k. A simple example of this group is the *translation group* $T(\mathbb{Z}_2^k)$ consisting of all permutations σ_y, $y \in \mathbb{Z}_2^k$, such that $\sigma_y(x) = x + y$. However, other maximal groups of involutions exist [306].

Take $f : \mathbb{Z}_2^k \to \mathbb{Z}_2^r$ and a maximal group G of involutions of $\mathbb{Z}_2^k$. Refer as the *generalized derivative of f in direction $\sigma \in G$* to the function $D_\sigma f(x) = f(\sigma(x)) - f(x)$. Note that if G is the translation group $T(\mathbb{Z}_2^k)$, then the generalized derivative coincides with the ordinary derivative $D_y f(x) = f(x + y) - f(x)$.

Definition 9 (Poinsot and Harari [306]). Take $A = (\mathbb{Z}_2^k, +)$, $B = (\mathbb{Z}_2^r, +)$, and a maximal group G of involutions of A. A function $f : A \to B$ is called a *G-bent function* if the generalized derivatives $D_\sigma f(x)$ in every direction $\sigma \in G$, $\sigma \neq e$, are balanced.

In the interpretation of Solodovnikov, a G-bent function is a function f that can be changed by every permutation $\sigma \in G$, $\sigma \neq e$, as strongly as possible: $\delta(f, f \circ \sigma) = 0$.

A translation of the definition of G-bent functions into the language of generalized Fourier coefficients is proposed in [306], but it fails to appear thoroughly worked out and includes inaccuracies. Also it is unclear how the approach in [306] can be applied if A and B are arbitrary Abelian groups.

15.11 MULTIDIMENSIONAL BENT FUNCTIONS ON A FINITE ABELIAN GROUP

In 2005, Poinsot [304] proposed this direct generalization of the bent functions of Logachev et al. [244].

Take the m-dimensional Hermitian space $\mathbb{C}^m$ with the standard inner product $\langle x, y \rangle = \sum_{j=1}^m x_j \bar{y}_j$, the norm $\|x\|^2 = \langle x, x \rangle$, and the metric $d(x, y) = \|y - x\|$. Suppose that $S_1(\mathbb{C}^m)$ is the set of all points lying on the sphere of radius 1 centered at the origin.

As before, take a finite Abelian group A of order n and its character group $\hat{A} = \{\chi_y | y \in A\}$.

Refer as the *Fourier transform* of a function $f : A \to \mathbb{C}^m$ to the function

$$\hat{f}(y) = \sum_{x \in A} f(x) \overline{\chi_y(x)}, \quad \hat{f} : A \to \mathbb{C}^m.$$

Definition 10 (Poinsot [304]). Take a finite Abelian group A of order n. A function $f : A \to S_1(\mathbb{C}^m)$ is called a *multidimensional bent function* if $\|\hat{f}(y)\|^2 = n$ for every $y \in A$.

For $m = 1$, this definition is identical to Definition 7. Similarly, for multidimensional bent functions, we have Rothaus's criterion, the dual multidimensional bent function can be defined, and so on [304]. However, so far it is unclear whether multidimensional bent functions can be of independent interest or are rather a formal generalization of the bent functions of Definition 7.

Some other ideas on multidimensional bent functions can be found in a paper by Nyberg and Hermelin [291].

CHAPTER 16

Combinatorial Generalizations of Bent Functions

INTRODUCTION

In this chapter we consider generalizations or subclasses based on some combinatorial ideas. We consider bent functions with some special properties—for example, with items of algebraic normal form being of the same degree (homogeneous functions), or highly symmetric bent functions (symmetric and rotation-symmetric bent functions) or bent functions that are constant/affine on subspaces (normal/weakly normal functions). We also consider non-Boolean functions similar to bent functions. In the sphere of our interest, there are symmetric bent functions, homogeneous bent functions, rotation-symmetric bent functions, normal bent functions, self-dual and anti-self-dual bent functions, partially defined bent functions, plateaued functions, $\mathbb{Z}$-bent functions and negabent functions.

16.1 SYMMETRIC BENT FUNCTIONS

A Boolean function f in n variables is called *symmetric* if for any permutation π on its coordinates $f(x) = f(\pi(x))$. It is the strongest symmetric property of a Boolean function. One can easily obtain that there are exactly 2^{n+1} symmetric Boolean functions since the value $f(x)$ depends only on the Hamming weight of x.

In 1994 Savicky [321] classified all symmetric bent functions.

Theorem 116. *There are only four symmetric bent functions in n variables:* $f(x), f(x) \oplus 1, f(x) \oplus \sum_{i=1}^{n} x_i$ *and* $f(x) \oplus \sum_{i=1}^{n} x_i \oplus 1$, *where*

$$f(x) = \bigoplus_{i=1}^{n} \bigoplus_{j=i+1}^{n} x_i x_j,$$

In 2006 Zhao and Li [398] discussed bent functions that have symmetric properties with respect to some variables.

Bent Functions
http://dx.doi.org/10.1016/B978-0-12-802318-1.00016-9

16.2 HOMOGENEOUS BENT FUNCTIONS

This subclass of bent functions was introduced by Qu et al. [313] in 2000 as consisting of the functions with relatively simple algebraic normal forms.

Definition 11 (Qu et al. [313]). A bent function is called *homogeneous* if all monomials of its algebraic normal form are of the same degree.

Qu et al. [313] enumerated all homogeneous bent functions of degree 3 in six variables (there are exactly 30 of them) and proposed the question of classifying the bent functions of this type with more variables. In 2002 Charnes et al. [77] proved that there are homogeneous bent functions of degree 3 in each number of variables $n > 2$.

In 2004 Xia et al. [382] established that, for $n > 3$, there are no homogeneous bent functions in n variables of the maximal possible degree $n/2$.

In 2007 Wang et al. [373] proposed a simple method to find all 30 homogeneous bent functions of degree 3 in six variables. Another proof of the nonexistence of homogeneous bent functions of degree $n/2$ in n variables (without using results from the difference set theory) was proposed.

Also in 2007, Meng et al. [266] generalized the results in [382]. They proved that for any nonnegative integer k, there is a positive integer N such that for $n \geqslant 2N$ there are no n-variable homogeneous bent functions having degree $(n/2) - k$ or more, where N is the smallest integer satisfying

$$2^{N-1} > \binom{N+1}{0} + \binom{N+1}{1} + \cdots + \binom{N+1}{k+1}.$$

But what is the exact upper bound on the degree of a homogeneous bent function? Presently there is no answer to this question. There is only a conjecture [266] that, for every $k > 1$, there is $N \geqslant 2$ such that homogeneous bent functions of degree k of n variables exist for every $n > N$.

In 2010 Meng et al. [263] presented partial results toward the conjectured nonexistence of homogeneous rotation-symmetric bent functions having degree more than 2.

16.3 ROTATION-SYMMETRIC BENT FUNCTIONS

In 1999 Pieprzyk and Qu [303] introduced a new concept of the rotation-symmetric Boolean function and applied it in the study of hash functions. One year earlier, Filiol and Fountain [120] discussed similar objects in terms of idempotents. But as mentioned in [96], rotation-symmetric Boolean functions first appeared in 1994 in a paper by Pieprzyk [302] in connection with bent permutations.

Let ρ be a cyclic permutation on coordinates $x_1, \ldots, x_n$ defined as

$$\rho(x_1, \ldots, x_{n-1}, x_n) = (x_n, x_1, \ldots, x_{n-1}) \text{ for all } x.$$

A Boolean function f in n variables is *rotation symmetric* if

$$f(x) = f(\rho(x)) \text{ for all } x \in \mathbb{F}_2^n.$$

Since the sum of rotation-symmetric Boolean functions is again rotation symmetric, it is possible to classify separately such homogeneous functions of degree k, where k runs through $\{0, 1, \ldots, n\}$, in order to get a full classification of rotation-symmetric Boolean functions.

Rotation-symmetric Boolean functions are widely studied all over the world. We recommend an excellent survey of them in a book by Cusick and Stănică [96]. There are several useful techniques for working with rotation-symmetric Boolean functions (like the *short algebraic normal form, SANF*), some classifications of rotation-symmetric functions, results on computing the number of them in distinct subclasses (usually based on Burnside's lemma), etc. Details of these and many other results related to rotation-symmetric Boolean functions can be found in papers by Stănică and Maitra [339, 340], Stănică et al. [341], and Clark et al. [90].

We briefly discuss results on rotation-symmetric bent functions. Classification of them for small n was done by Stănică and Maitra [339, 340] in 2003 and 2008.

If $n = 4$, there are eight rotation-symmetric bent functions in four variables. Their SANFs (up to a linear part) are 13 and $12 + 13$.

If $n = 6$, there are 48 rotation-symmetric bent functions in n variables. All of them can be represented by the following 12 functions (free of linear terms): 14, $12 + 13 + 14$, $134 + 13 + 14$, $124 + 13 + 14$, $124 + 12 + 14$, $134 + 12 + 14$, $123 + 135 + 14$, $123 + 135 + 12 + 13 + 14$, $123 + 134 + 135 + 13 + 14$, $123 + 134 + 135 + 12 + 14$, $123 + 124 + 135 + 12 + 14$, and $123 + 124 + 135 + 13 + 14$. We list them here in SANF. This means that in order to obtain the classical algebraic normal form of a function, we should take all cyclic shifts of the SANF. For example, a bent function with SANF 14 is equal to $14 + 25 + 36$, or in the full form is equal to $x_1 x_4 \oplus x_2 x_5 \oplus x_3 x_6$. Then to get 48 rotation-symmetric functions in six variables we add a rotation-symmetric affine part of four types: zero, one, $x_1 \oplus \cdots \oplus x_n$, or $x_1 \oplus \cdots \oplus x_n \oplus 1$.

Stănică and Maitra found among the 2^{21} rotation-symmetric Boolean functions in eight variables that exactly 15 104 of them are bent functions. There are exactly eight homogeneous rotation-symmetric bent functions in eight variables: 15, $15 + 12$, $15 + 13$, $15 + 14$, $15 + 12 + 13$, $15 + 12 + 14$, $15 + 13 + 14$, and $15 + 12 + 13 + 14$. We list them again in SANF. It is easy to see some *group structure* in this construction. Indeed, let us take some basis of an Abelian group G isomorphic to $\mathbb{Z}_2^3$; denote basic vectors by formal symbols "12," "13," and "14." Then the SANFs of all homogeneous bent functions in eight variables are exactly elements of the set "15" $+ G$.

Stănică and Maitra [340] also studied the case $n = 10$. But it is still difficult to classify all rotation-symmetric bent functions in 10 variables. It was found that there are 12 homogeneous rotation-symmetric bent functions in 10 variables of degree 2. They are as follows: 16, $16 + 12$, $16 + 13$, $16 + 14$, $16 + 15$, $16 + 12 + 15$, $16 + 13 + 14$, $16 + 12 + 13 + 14$, $16 + 12 + 13 + 15$, $16 + 12 + 14 + 15$, $16 + 13 + 14 + 15$, and $16 + 12 + 13 + 14 + 15$. Stănică and Maitra [340] did not find homogeneous rotation-symmetric bent functions of greater degree. Then, they proposed a conjecture: there are no homogeneous rotation-symmetric bent functions of degree 3 or more.

There is a some progress in proving this conjecture in work by Stănică [336]. Further, we follow to this work and [96].

For a homogeneous degree d rotation-symmetric Boolean function f with its SANF given by $\bigoplus_{i=1}^{s} \beta_i$, where $\beta_i = x_{k_1^{(i)}} x_{k_2^{(i)}} \cdots x_{k_d^{(i)}}$ (assume that $k_1^{(i)} = 1$ for all i), we define a sequence $d_j^{(i)}$, $j = 1, 2, \ldots, k_{i-1}^{(i)}$, by $d_j^{(i)} = k_{j+1}^{(i)} - k_j^{(i)}$. Let $d_f = \max_{i,j}\{d_j^{(i)}\}$—that is, the largest distance between two consecutive indices in all monomials of f. The next theorem was proven by Stănică [336] in 2008, and it generalizes in some direction the results presented in [95, 302]:

Theorem 117. *The following hold for a homogeneous rotation-symmetric Boolean function f of degree 3 or greater in $n \geqslant 6$ variables:*

(1) *If the SANF of f is $x_1 \cdots x_d$, then f is not a bent function.*

(2) *If the SANF of f is $x_1 x_2 \cdots x_{d-1} x_d \oplus x_1 x_2 \cdots x_{d-1} x_{d+1}$, then f is not bent, assuming $(n-2)/4 > \lfloor n/d \rfloor$, if $n \neq 1 \pmod{d}$; $n/4 > \lfloor n/d \rfloor$, if $n = 1 \pmod{d}$.*

(3) *In general, if $d_f < (n/2 - 1)/\lfloor n/d \rfloor$, then f is not bent.*

In 2009 Dalai et al. [99] analyzed combinatorial properties related to the Walsh-Hadamard spectra of rotation-symmetric Boolean functions in an even number of variables. These results were then applied in the study of rotation-symmetric bent functions.

Constructions of quadratic and cubic rotation-symmetric bent functions can be found in a paper by Gao et al. [134]. For example, they construct the first infinite class of cubic rotation-symmetric bent functions.

In 2014 new constructions of rotation-symmetric bent functions via idempotents were proposed by Carlet et al. [59]. Namely, they found the first infinite class of such functions of degree more than 3.

16.4 NORMAL BENT FUNCTIONS

Recall that a Boolean function f in n variables is *affine on a set* $L \subseteq \mathbb{F}_2^n$ if there is a vector $a \in \mathbb{F}_2^n$ and a constant $b \in \mathbb{F}_2$ such that $f(x) = \langle a, x \rangle \oplus b$ for every $x \in L$. A Boolean function is *constant on a set* $L \subseteq \mathbb{F}_2^n$ if there is a constant $b \in \mathbb{F}_2$ such that $f(x) = b$ for all $x \in L$.

In 1995 Dobbertin [111] introduced the following important notion:

Definition 12 (Dobbertin [111]). A Boolean function f in n variables is called *normal* (*weakly normal*) if there is an $(n/2)$-dimensional subspace (flat) of the space $\mathbb{F}_2^n$ such that f is constant (or affine) on it.

Dobbertin proposed such a notion for constructing highly nonlinear balanced functions. Namely, he used it to construct class $\mathcal{N}$ of normal bent functions on the basis of the following fact: if a bent function f in n variables is constant on some affine $n/2$-dimensional subspace L, then f is balanced on any other of its cosets that does not coincide with L.

All Maiorana McFarland bent functions are normal (remember Theorem 36), and all bent functions from $\mathcal{PS}^+$ and $\mathcal{PS}_{ap}$ are normal too. Whether the functions from $\mathcal{PS}^-$ are normal or weakly normal is still unknown.

For 10 years—from 1995 to 2005—there was a hypothesis that every bent function is normal. In 2006 this hypothesis was disproved: Canteaut et al. [33] found examples of nonnormal and non-weakly normal bent functions. For example, all Kasami bent functions in six variables are normal, but in [33] it is shown that

- if $n = 10$, then for $\alpha \in \mathbb{F}_4 \setminus \mathbb{F}_2 \subset \mathbb{F}_{2^{10}}$ we have $\beta \in \mathbb{F}_{2^{10}}$ such that the bent function f in n variables defined as $f(x) = \text{tr}(\alpha x^{57} \oplus \beta x)$ is nonnormal;
- if $n = 14$, a Kasami bent function $f(c) = \text{tr}(\alpha c^{57})$, where $\alpha \in \mathbb{F}_4 \setminus \mathbb{F}_2 \subset \mathbb{F}_{2^{10}}$, is nonweakly normal (and hence nonnormal too).

These two statements were verified using an algorithm introduced by Daum et al. [101]. Note that any Kasami bent function in n variables is normal if 6 divides n; see [33].

Next, there are several iterative constructions saving normality (weak normality) and nonnormality (nonweak normality).

Canteaut et al. [33] proved the following theorem:

Theorem 118. *Let f be a Boolean function in n variables. Then f is normal (weakly normal) if and only if the function g in $n+2$ variables defined by $g(x, x_{n+1}, x_{n+2}) = f(x) \oplus x_{n+1}x_{n+2}$ is normal (weakly normal).*

In 2004 Charpin [78] generalized the notion of normality by introducing k-normality. A Boolean function is called *k-normal (weakly k-normal)* if it is constant (affine) on an affine subspace of dimension k. Gangopadhyay and Sharma [131] obtained the following result:

Theorem 119. *Let f_1 and f_2 be Boolean functions in n variables. Then f_1 or f_2 is weakly k-normal if and only if the Boolean function g in $n+2$ variables given by $g(x, x_{n+1}, x_{n+2}) = f_1(x) \oplus x_{n+1}x_{n+2} \oplus (x_{n+1} \oplus x_{n+2})(f_1(x) \oplus f_2(x))$ is weakly $(k+1)$-normal.*

A notion analogous to k-normality was studied in 1984 by Ryabov [319] and later in 2005 by Buryakov and Logachev [28]. In their works, the bounds on the maximal possible dimension of a subspace on which a function is constant (affine) were proposed. Their bounds work for almost all Boolean functions.

Carlet et al. [56] proved that if in the simplest iterative construction (Theorem 32) one bent function g is normal and the other h is not, then the resulting function $g \oplus h$ is a nonnormal bent function. This was proven using a notion of *normal extension* of a bent function.

Thus, starting with $n = 10$ ($n = 14$) variables, one can construct non-normal (non-weakly normal) bent functions for any even n.

The notion of normality plays an important role in the study of metrical properties of bent functions; see Chapter 11 and papers by Kolomeec and Pavlov [207] and Kolomeec [208, 210].

16.5 SELF-DUAL AND ANTI-SELF-DUAL BENT FUNCTIONS

Recall that for a bent function f the *dual function* $\tilde{f}$ in n variables is defined by the equality

$$W_f(y) = 2^{n/2}(-1)^{\tilde{f}(y)}.$$

This definition is correct since $W_f(y) = \pm 2^{n/2}$ for any vector y. It is easy to see that the function $\tilde{f}$ is bent too. It holds that $\tilde{\tilde{f}} = f$. Some relations between a bent function and its dual were considered in Section 5.4.

Of special interest are *self-dual bent functions* (such that $f = \tilde{f}$) and *anti-self-dual bent functions* (such that $f = \tilde{f} \oplus 1$).

In 2008 and 2010 several steps in their classification were made by Carlet et al. [52]. Namely, they gave a spectral characterization of the sign functions of these two classes of functions in terms of the Rayleigh quotient of the Sylvester Hadamard matrix. Namely, the *Rayleigh quotient* S_f of a Boolean function f in n variables is the sum

$$S_f = \sum_{x,y \in \mathbb{F}_2^n} (-1)^{f(x) \oplus f(y) \oplus \langle x,y \rangle}.$$

As mentioned by Danielsen et al. [100], the Rayleigh quotient of a bent function is an invariant under the action of the orthogonal group, and it measures the distance of the function to its dual.

The following theorem is proven in [52]:

Theorem 120. *Let f be a Boolean function in n variables (n is even). The modulus of the character sum S_f is at most $2^{3n/2}$ with equality if and only if f is self-dual bent or anti-self-dual bent.*

This spectral characterization allowed Carlet et al. [52] to give a simple and efficient search algorithm that makes it possible to enumerate and classify all self-dual bent function for fewer than six variables and all such quadratic functions in eight variables. Carlet et al. [52] proposed characterizations of self-dual bent functions in Maiorana-McFarland (Theorem 38) and Dillon (Theorem 41) classes; they studied the self-duality in iterative constructions of bent functions. Operations on Boolean functions that preserve self-duality or anti-self-duality were also investigated.

An interesting corollary of their results related to metrical properties is that the Hamming distance between a self-dual bent function and an anti-self-dual bent function, both in n variables, is 2^{n-1}; see [52].

In 2012 Hou [173] classified all self-dual and anti-self-dual quadratic bent functions in n variables under the action of the orthogonal group $O(n, \mathbb{F}_2)$. This was done through a classification of all $n \times n$ involutory alternating matrices over $\mathbb{F}_2$ under the action of the orthogonal group. The sizes of the $O(n, \mathbb{F}_2)$-orbits of self-dual and anti-self-dual quadratic bent functions were determined explicitly. In 2013 Hou [174] proposed a classification of self-dual quadratic p-ary bent functions when p is an odd prime.

In 2012 Hyun et al. [180] proved that the MacWilliams duality holds for bent functions. Then the concept of formally self-dual Boolean functions with respect to their near weight enumerators was proposed. By using this concept, Hyun et al. proved the Gleason-type theorem on self-dual bent functions and provided the total number of (self-dual) bent functions in two and four variables obtained from formally self-dual Boolean functions.

In 2013 Feulner et al. [119] classified self-dual quadratic and cubic bent functions in eight variables via a computer search. There are exactly 4 and 45 nonequivalent self-dual bent functions of degree 2 and 3, respectively. This result was obtained by enumerating all eigenvectors with ± 1 entries of the Sylvester Hadamard matrix with an integer programming algorithm based on lattice basis reduction.

Bent functions with respect to their duals have been intensively studied by Mesnager. In 2014 she proposed new constructions of bent functions, self-dual bent functions, and anti-self-dual bent functions [274].

16.6 PARTIALLY DEFINED BENT FUNCTIONS

Given an arbitrary set $S \subseteq \mathbb{F}_2^n$, take a *partially defined* Boolean function $f : S \to \mathbb{F}_2$. Its *partial Walsh-Hadamard transform* is the mapping

$$W_{f,S}(\gamma) = \sum_{x \in S} (-1)^{\langle x, \gamma \rangle \oplus f(x)} \quad \text{for every} \quad \gamma \in \mathbb{F}_2^n.$$

This transformation satisfies the analogue of Parseval's equality:

$$\sum_{\gamma \in \mathbb{F}_2^n} W_{f,S}^2(\gamma) = 2^n |S|.$$

Definition 13. A Boolean function f is called a *partially defined bent function* if $W_{f,S}(\gamma) = \pm\sqrt{|S|}$ for every $\gamma \in \mathbb{F}_2^n$.

These functions are discussed in more detail in [247, chapter 6]. Here we note only that it is unknown under what conditions on S partially defined bent functions exist.

16.7 PLATEAUED FUNCTIONS

This generalization of bent functions is well known, and our discussion is very brief here.

Definition 14. A Boolean function is called *plateaued* if all its nonzero Walsh-Hadamard coefficients are equal in absolute value.

Parseval's equality implies that the nonzero coefficients must be of the form $\pm 2^{n-h}$ for some integer h with $0 \leqslant h \leqslant n$. The number of nonzero coefficients must be equal to 2^{2h}. The exponent $2h$ and the quantity 2^{n-h} are called, respectively, the *order* and the *amplitude* of a plateaued function. The bent functions and the affine functions are the marginal particular cases of the plateaued functions (of orders n and 0, respectively).

For results on these functions, see the surveys [46, 247], as well as [67, 403, 404].

The question of how to construct bent functions from plateaued functions was discussed in 2013 by Çeşmelioğlu and Meidl [75].

16.8 $\mathbb{Z}$-BENT FUNCTIONS

In 2005 Dobbertin and Leander [113] suggested studying bent functions in the context of a more general approach, which we can call *recursive*. We do not distinguish between an ordinary Boolean function $f(x)$ of $x \in \mathbb{F}_2^n$ and the integer function $F(x) = (-1)^{f(x)}$. The *Fourier transform* of a function $F : \mathbb{F}_2^n \to \mathbb{Z}$ is defined as

$$\hat{F}(y) = \frac{1}{2^{n/2}} \sum_{x \in \mathbb{F}_2^n} (-1)^{\langle x, y \rangle} F(x).$$

Then a (± 1)-valued function F is a bent function if and only if $\hat{F}$ is (± 1)-valued as well. The generalization is as follows:

Definition 15 (Dobbertin and Leander [113]). Given $T \subseteq \mathbb{Z}$, a function $F : \mathbb{F}_2^n \to T$ is called a *T-bent function* if all values of $\hat{F}$ belong to T.

Dobbertin and Leander chose the natural nested chain

$$W_0 = \{-1, +1\};$$
$$W_r = \{w \in \mathbb{Z} | -2^{r-1} \leqslant w \leqslant 2^{r-1}\}, \quad r > 0.$$

A W_r-bent function is called a $\mathbb{Z}$-*bent function of level* r, and all these bent functions (for $r \in \mathbb{Z}$, $r \geqslant 0$) constitute the class of $\mathbb{Z}$-*bent functions*. For more details on this theme please see a paper by Dobbertin and Leander [114] from 2008. It is shown there how $\mathbb{Z}$-bent functions of lower level can be built up recursively by gluing together $\mathbb{Z}$-bent functions of higher level. This recursion stops at level zero, containing the usual bent functions. In [114] bent functions are studied in the framework of $\mathbb{Z}$-bent functions. and some guidelines for further research are given.

In 2013 Gangopadhyay et al. [128] generalized the construction of partial spread bent functions (Theorem 39) to partial spread $\mathbb{Z}$-bent functions of an arbitrary level.

Theorem 121. *If $m_1, m_2, \ldots, m_s \in \mathbb{Z}$ and $E_1, E_2, \ldots, E_s$ are k-dimensional subspaces of $\mathbb{F}_2^n$, then the function $f(x) = \sum_{i=1}^{s} m_i \mathrm{Ind}_{E_i}(x)$ is a $\mathbb{Z}$-bent function and its dual is given by $\sum_{i=1}^{s} m_i \mathrm{Ind}_{E_i^\perp}(x)$.*

Theorem 122. *Suppose $\{E_i : i = 1, 2, \ldots, s\}$ is a set of k-dimensional subspaces of $\mathbb{F}_2^n$ with the property that $E_i \cap E_j = \{0\}$ whenever $i \neq j$. The function $f(x) = \sum_{i=1}^{s} m_i \mathrm{Ind}_{E_i}(x)$, where $m_i \in W_r$, for all $i = 1, 2, \ldots, s$, is a $\mathbb{Z}$-bent function of level r, for any $r \geqslant 1$, if and only if $\sum_{i=1}^{s} m_i \in W_r$.*

Furthermore, Gangopadhyay et al. showed how these partial spread $\mathbb{Z}$-bent functions give rise to a new construction of (classical) bent functions. Also they constructed a bent function in eight variables which is nonequivalent to all Maiorana-McFarland bent functions as well as $\mathcal{PS}_{\mathrm{ap}}$-type bent functions. All bent functions in six variables, up to equivalence, can be obtained by their construction [128].

16.9 NEGABENT FUNCTIONS, BENT$_4$-FUNCTIONS, AND *I*-BENT FUNCTIONS

A bent function is often defined as a function with a *flat spectrum* of the Walsh-Hadamard transform. Flatness means that the absolute values of all Walsh-Hadamard coefficients are equal. In 2006, Riera and Parker [314] began to study Boolean functions with flat spectra of a set of unitary transformations of a particular form. Recall that the transformation of the space $\mathbb{C}^n$ given by a square matrix A is *unitary* if $A\bar{A}^{\mathrm{T}} = E$, where E is the identity matrix. These transformations are used in [45] to analyze the stabilizers of quantum states. Put

$$H = \frac{1}{\sqrt{2}} \begin{pmatrix} 1 & 1 \\ 1 & -1 \end{pmatrix}, \qquad I = \begin{pmatrix} 1 & 0 \\ 0 & 1 \end{pmatrix}, \qquad N = \frac{1}{\sqrt{2}} \begin{pmatrix} 1 & i \\ 1 & -i \end{pmatrix}.$$

For every 2×2 matrix A, let $A_j = I \otimes \cdots \otimes I \otimes A \otimes I \otimes \cdots \otimes I$ denote the tensor (Kronecker) product of n matrices, where A appears in position j. Consider the following sets of transformations:

- $\{H\}^n$ consisting of the transformations $U = \prod_{j=0}^{n-1} H_j$. If $F = (-1)^f$ is the *sign* function of a Boolean function f of n variables, then the vector of spectral values of f with respect to the transformation U is defined as $\hat{F} = UF$. Then f is a *bent function* (in the usual sense) if its spectrum with respect to U is flat—that is, every component of $\hat{F}$ is equal to ± 1.

- $\{N\}^n$ consisting of the transformations $U = \prod_{j=0}^{n-1} N_j$.

Definition 16 (Riera and Parker [314]). A Boolean function having a flat spectrum with respect to U is called a *negabent function*.

Note that since U is a complex matrix, the definition of a spectrum here involves certain specific features [297]. Every affine Boolean function is a negabent function. Parker [296] and Parker and Pott [297] studied negabent functions. The intersection of the classes of bent and negabent functions is considered in [297]; it is completely understood for quadratic functions:

- $\{H, N\}^n$ consisting of 2^n transformations of the form $\prod_{j \in R_H} H_j \prod_{j \in R_N} N_j$, where R_H and R_N partition the set $\{0, 1, \ldots, n-1\}$. A Boolean function f of n variables is a *bent$_4$-function* if there is at least one partition R_H, R_N for which the spectrum of f is flat.

- $\{I, H\}^n$ consisting of 2^n transformations of the form $\prod_{j \in R_I} I_j \prod_{j \in R_H} H_j$, where R_I and R_H partition the set $\{0, 1, \ldots, n-1\}$. By analogy to the previous case, a function f is an *I-bent function* if there is at least one partition R_I, R_H with $|R_I| < n$ for which the spectrum of f is flat.

- $\{I, H, N\}^n$ consisting of 3^n transformations $\prod_{j \in R_I} I_j \prod_{j \in R_H} H_j \prod_{j \in R_N} N_j$, where R_I, R_H, and R_N partition $\{0, 1, \ldots, n-1\}$. In this case we may define the so-called *I-bent$_4$-functions*, which, however, are of little interest since this class includes all Boolean functions.

In 2006 Riera and Parker [314] developed the quantum direction of their research, and studied the properties of the bent functions of the new type and their connections to graphs. In 2010 and 2012 Stănică et al. [337, 338] investigated bent and negabent functions via nega-Hadamard transform. In 2013 Su et al. [345] proposed a characterization of negabent functions with maximal algebraic degree. In 2013 Gangopadhyay et al. [130] obtained a relationship between bent, semibent, and bent$_4$-functions which is a generalization of the relationship between bent and negabent Boolean functions proved by Parker and Pott [297]. As a corollary, they proved that the maximal possible algebraic degree of a bent$_4$-function in n variables is $[n/2]$, and hence they solved an open problem posed by Riera and Parker [314].

CHAPTER 17

Cryptographic Generalizations of Bent Functions

INTRODUCTION

It is known that for a Boolean function to be "good" from the cryptographic point of view it should possess several cryptographic properties. Often such properties contradict each other. The problem of constructing bent functions with stronger cryptographic properties is again considered in this chapter. Here we concentrate on additional restrictions for bent functions that lead to new cryptographic notions (such as balanced bent functions and hyperbent functions) and also on modifications of bent functions by the following principle: "less nonlinearity, stronger other properties" (partially bent functions). We consider semibent functions, balanced (semi-) bent functions, partially bent functions, hyperbent functions, bent functions of higher order, and k-bent functions.

17.1 SEMIBENT FUNCTIONS (NEAR-BENT FUNCTIONS)

This is a very natural generalization of a bent function in the case of an odd number of variables.

Definition 17. A Boolean function f is called a *semibent function* if $W_f(\gamma)$ equals 0 or $2^{(n+1)/2}$ for every $\gamma \in \mathbb{F}_2^n$.

In the case of p-ary semibent functions, all the Walsh-Hadamard coefficients should be equal to 0 or $p^{(n+1)/2}$. This definition was introduced by Chee et al. [83] in 1994. In the literature there are several names for semibent functions. Sometimes these functions are called *near-bent functions* or *three-valued almost optimal Boolean functions*, but the term *semibent functions* is more widespread.

It is clear that semibent functions are nothing but the plateaued functions of maximal order $n - 1$ in an odd number of variables; see Definition 14. Such functions with three distinct values in the Walsh-Hadamard spectrum are interesting for defending against the so-called soft output joint attack on pseudorandom number generators [233], which are used in

Bent Functions
http://dx.doi.org/10.1016/B978-0-12-802318-1.00017-0

the IS-95 standard of code division multiple access (CDMA) technology. Semibent functions are used for constructing the cryptographically robust S-blocks [106].

It is known that balanced semibent functions exist (we consider them in Section 17.2). This fact (together with the possibility to combine them with other cryptographical properties) makes semibent functions very attractive for cryptographic applications. This is one of the most intensively studied topics related to bent functions; and of course it requires a separate investigation.

We give here only a very brief and noncomprehensive overview of some recent publications on this theme.

In 2005 Charpin et al. [81] showed the connection between bent and semibent functions. In particular, they showed how to concatenate two suitably chosen semibent functions to get a bent function and vice versa. A generalization of this approach was also considered.

In 2006 a characterization of semibent functions over finite fields was proposed by Khoo et al. [197].

In 2008 Dillon and McGuire [110] presented a general criterion for semibent functions to be bent on a hyperplane. One year later, Leander and McGuire [232] proposed a construction of bent functions in dimension $2m$ from semibent functions in dimension $2m - 1$. In particular, they constructed the first examples of non-weakly normal bent functions in dimensions 10 and 12 in order to demonstrate the significance of their construction. In 2009 Sun and Wu [346] considered the question of constructing semibent functions in an even number of variables. In 2011 autocorrelation coefficients of semibent functions were studied by Li et al. [238]. He and Ma [154] investigated balanced semibent functions with high algebraic degrees.

In 2012 Carlet and Mesnager [66] proposed an interesting approach for constructing semibent functions from bent functions. They proved that a Boolean function, in an even number of variables, equal to the sum of a Boolean function g which is constant on each element of a spread and of a Boolean function h whose restrictions to these elements are all linear, is semibent if and only if functions g and h are both bent. A large number of semibent functions (for any n) were constructed in the explicit polynomial form.

Semibent functions with Niho exponents were considered in 2012 by He et al. [153]. In 2013 Dong et al. [116] proposed new constructions of semibent functions in polynomial forms; see also a paper by Chen and Cao

[85]. In 2013 a new class of quadratic semibent Boolean functions with some special trace form was proposed by Tang et al. [349].

Spreading codes for CDMA constructed from semibent functions were considered in 2012 by Hunt and Smith [178].

In 2013 Mesnager [273] showed how oval polynomials can be applied in the construction of semibent functions in an even number of variables. It is one of the most interesting examples of how the objects related to finite projective geometry (such as hyperovals) can be linked to cryptographic Boolean functions. We recommend this approach for further deep study.

In 2014 Zheng et al. [400] considered quadratic bent and semibent functions over finite fields of odd characteristic.

17.2 BALANCED (SEMI-) BENT FUNCTIONS

From the viewpoint of cryptography, the important criteria a Boolean function f in n variables must satisfy are as follows [46, 246]:

- *Balancedness*, which means that f takes values 0 and 1 equally often.
- *Order k propagation criterion* PC(k), which means that, for every nonzero vector $y \in \mathbb{F}_2^n$ of weight at most k, where $1 \leqslant k \leqslant n$, the function $f(x \oplus y) \oplus f(x)$ is balanced [309].
- *Maximal nonlinearity*, which means that f is such that the value of its nonlinearity N_f is maximal.
- *Uniform correlation with linear functions*. The correlation between two functions f and g is defined as $c(f,g) = 1 - \frac{\text{dist}(f,g)}{2^{n-1}}$; for a function f the uniform correlation means that the value of $|c(f,g)|$ is constant for every linear function g.

However, these criteria contradict each other. Bent functions are maximally nonlinear, satisfy the criterion PC(n), possess uniform correlation with linear functions (the value is equal to $\pm 2^{-n/2}$), but are never balanced. The following definition arises naturally:

Definition 18. A Boolean function f in n variables is called a *balanced bent function* if f is balanced and has the maximal possible nonlinearity.

It was established in [3] that if n is odd and f is a balanced function, then $N_f \leqslant 2^{n-1} - 2^{(n-1)/2}$.

In 1994, Chee et al. [83] proposed a method for constructing the balanced bent functions of odd numbers of variables possessing almost uniform correlation with the linear functions and satisfying the criterion PC(k) for sufficiently large k. We present this method.

Given some odd n, take a nonsingular binary $(n-1) \times (n-1)$ matrix A and a binary vector b of length $n-1$.

Theorem 124. *If f_0 is a bent function in $n-1$ variables and f_1 is the equivalent bent function $f_1(x) = f_0(Ax + b) + 1$, then the function $g(x, z) = f_z(x)$ in n variables, where $x \in \mathbb{F}_2^{n-1}$ and $z \in \mathbb{F}_2$,*

(1) *is a balanced bent function;*

(2) *is a balanced semibent function (see Section 17.1);*

(3) *has only the possible values 0 and $\pm 2^{-(n-1)/2}$ of correlation with a linear function;*

(4) *satisfies the propagation criterion for every nonzero vector $(y, 0)$, where $y \in \mathbb{F}_2^{n-1}$;*

(5) *satisfies the PC$(n-1)$ if $A = E$ and b is a vector of all 1's.*

Another approach to combine "balancedness" and "nonlinearity" was proposed by Zhang et al. [393] in 2011. They defined a class of *nearly bent functions*. These functions lie at a large Hamming distance to all affine functions and preserve a high level of algebraic degree. A construction of near-bent functions was described. It was shown that the constructed functions have good global avalanche characteristic properties.

17.3 PARTIALLY BENT FUNCTIONS

As we have already noted, bent functions are neither balanced nor correlation immune. Carlet [35] proposed a new method to extend the class $\mathcal{B}_n$ to functions having these properties and having sufficiently high nonlinearity. These *partially bent functions* are defined using the following extremal property:

Denote the *autocorrelation* of a Boolean function f in direction y by

$$\Delta_f(y) = \sum_{x \in \mathbb{F}_2^n} (-1)^{f(x) \oplus f(x \oplus y)}.$$

Let NW_f and $\mathrm{N}\Delta_f$ denote the numbers of nonzero Walsh–Hadamard coefficients and autocorrelation coefficients of f, respectively. Then every Boolean function satisfies $\mathrm{NW}_f \cdot \mathrm{N}\Delta_f \geqslant 2^n$ [35].

Definition 19 (Carlet [35]). A Boolean function f with $\mathrm{NW}_f \cdot \mathrm{N}\Delta_f = 2^n$ is called a *partially bent function*.

Theorem 125. *The following claims are equivalent:*

(1) *f is a partially bent function.*

(2) *There is a vector z such that, for every x, the value of the autocorrelation $\Delta_f(x)$ is equal to either 0 or $(-1)^{\langle x,z\rangle} 2^n$.*

(3) *There is a vector z and a decomposition of $\mathbb{F}_2^n$ into the direct sum of subspaces L and L' such that $f|_{L'}$ is a partially defined bent function (in the sense of Definition 8), and $f(x \oplus y) = \langle x, z\rangle \oplus f(y)$ for every $x \in L$ and $y \in L'$.*

Henceforth, z stands for the vector defined in Theorem 125. The subspace L for a partially bent function f is defined as the set of vectors x such that $\Delta_f(x) \neq 0$. We can equivalently define L as the space of linear structures of f—that is, the space consisting of all vectors y with $f(x \oplus y) \oplus f(x) = $ const. For a decomposition of $\mathbb{F}_2^n$ into a direct sum, the subspace L' is chosen arbitrarily. Note that the dimension of L' must be even, and denote it by $2h$. According to [35], the following theorem is true (for the necessary definitions, see [247]):

Theorem 126. *A partially bent function f is*

(1) *balanced if and only if $f|_L \neq$ const;*

(2) *unbalanced of weight w if and only if $f|_L$ is a constant and $w = 2^{n-1} \pm 2^{n-h-1}$, where $\dim L = n - 2h$;*

(3) *a plateaued function of order $2h$;*

(4) *a correlation-immune function of order k if and only if there is no vector of weight w, $1 \leqslant w \leqslant k$, in the dual class $z \oplus L^{\perp}$;*

(5) *a balanced correlation-immune function of order k if and only if there is no vector of weight at most k in the class $z \oplus L^{\perp}$;*

(6) *a function that satisfies the PC(k) if and only if L does not include any vector of weight w, $1 \leqslant w \leqslant k$.*

Note that all affine, quadratic, and bent functions are partially bent functions.

The following theorem holds [35]:

Theorem 127. *Let f be a partially bent function, $\dim L = n - 2h$. Then*

$$N_f = 2^{n-1} - 2^{n-h-1},$$

$$W_f(x) = \begin{cases} \pm 2^{n-h}, & \text{for } x \in z + L^{\perp}, \\ 0, & \text{otherwise.} \end{cases}$$

Obviously, the lower the dimension of the space L, the higher is the nonlinearity of a partially bent function.

For more on this topic, see [46, 374].

17.4 HYPERBENT FUNCTIONS

In 2001, Youssef and Gong [389] introduced the concept of hyperbent functions.[1] Previously, in 1999, Gong and Golomb [143] had considered the Data Encryption Standard (DES) ciphering algorithm as a nonlinear feedback shift register, and analyzed its S-blocks. For this approach, Gong and Golomb of [143] proposed using proper monomial functions instead of linear Boolean functions for approximating the coordinate functions of the S-blocks. This idea was developed in [389].

We can regard a Boolean function in n variables as a function from $\mathbb{F}_{2^n}$ to $\mathbb{F}_2$, assigning to every vector x a corresponding element of the field $\mathbb{F}_{2^n}$. It is known that every linear function $\langle x, y \rangle$ can be expressed as $\mathrm{tr}(a_x y)$ for suitable $a_x \in \mathbb{F}_{2^n}$, where $\mathrm{tr} : \mathbb{F}_{2^n} \to \mathbb{F}_2$ is the trace function. Then the Walsh-Hadamard transform assumes the equivalent form

$$W_f(y) = \sum_{x \in \mathbb{F}_{2^n}} (-1)^{\mathrm{tr}(yx)+f(x)}.$$

A function of the form $\mathrm{tr}(a_x y^s)$, where the integer s satisfies $1 \leqslant s \leqslant 2^n - 1$ and $\gcd(s, 2^n - 1) = 1$, is called a *proper monomial function*.

The *extended Walsh-Hadamard transform* of a Boolean function f is

$$W_{f,s}(y) = \sum_{x \in \mathbb{F}_{2^n}} (-1)^{\mathrm{tr}(yx^s)+f(x)}.$$

Definition 20 (Youssef and Gong [389]). A Boolean function f is called a *hyperbent function* if $|W_{f,s}(y)| = 2^{n/2}$ for every $y \in \mathbb{F}_{2^n}$ and every integer s with $\gcd(s, 2^n - 1) = 1$.

It is possible to say also that a bent function is hyperbent if it is bent up to a change of primitive roots in the finite field $\mathbb{F}_{2^n}$. In other words, a hyperbent function is equally badly approximated by all proper monomial functions; its generalized nonlinearity

$$\mathrm{NLG}(f) = 2^{n-1} - \frac{1}{2} \max_{y,s \in \{y,s | \gcd(s, 2^n - 1) = 1\}} |W_{f,s}(y)|$$

is maximal: it is equal to $2^{n-1} - 2^{(n/2)-1}$.

[1] Prior to that the term *hyperbent function* was used once, in [39], for another class of functions, but that sense is not used anymore.

For every even n, Youssef and Gong [389] proved the existence of hyperbent functions, proposed their vector version, and considered balanced hyperbent functions for small numbers of variables.

In 2006, Kuz'min et al. [220] and independently Carlet and Gaborit [58] showed that the degree of every hyperbent function in n variables is equal to $n/2$.

Theorem 128. *The degree of an arbitrary hyperbent function in n variables is equal to $n/2$.*

In 2007 and 2008 Kuz'min et al. [221, 222] generalized the concept of a hyperbent function: from Boolean functions they passed to functions over an arbitrary finite field of characteristic 2.

Take $q = 2^\ell$. The problem of approximating an arbitrary function from $\mathbb{F}_q^n$ to $\mathbb{F}_q$ (as above, it is identified with a function $f : \mathbb{F}_{q^n} \to \mathbb{F}_q$) by functions of some bounded class $\mathcal{A}$ is considered in [221]. To estimate the efficiency of approximation of f by a function $g \in \mathcal{A}$, the *agreement* parameter $\nabla(f,g)$ was related to the closeness function of Solodovnikov (15.4) as

$$\nabla(f,g) = \frac{q}{\sqrt{q-1}} \delta(f,g)$$

if we choose the finite groups $A = (\mathbb{F}_{q^n}, +)$ and $B = (\mathbb{F}_q, +)$. This parameter appears more natural since $0 \leqslant \nabla(f,g) \leqslant 1$; and, for the marginal values 0 and 1, the functions f and g differ by a balanced function and a constant, respectively. For $q = 2$, we have

$$\left| P(f = g) - \frac{1}{2} \right| = \frac{\nabla(f,g)}{2};$$

thus, the less agreement there is between two functions, the lower the efficiency of replacing one with the other.

Let

$$\nabla(f, \mathcal{A}) = \max_{g \in \mathcal{A}} \nabla(f,g)$$

denote the *efficiency of approximation* of f by functions from $\mathcal{A}$.

- If $\mathcal{A} = \mathrm{Hom}(A, B)$ is the class of all homomorphisms from A to B, then every function $f : \mathbb{F}_{q^n} \to \mathbb{F}_q$ such that $\nabla(f, \mathrm{Hom}(A, B))$ takes its minimal possible value $q^{-n/2}$ is a bent function in the sense of Definition 4.
- Suppose that $\mathcal{A} = \mathcal{M}$ is the class of all proper generalized monomial functions—that is, the functions of the form $g(x) = h(x^s)$, where $h \in \mathrm{Hom}(A, B)$ and the integer s satisfies $\gcd(s, q^n - 1) = 1$.

Definition 21 (Kuz'min et al. [222]). A function $f : \mathbb{F}_{q^n} \to \mathbb{F}_q$ is called a *hyperbent function over the field* if the parameter $\nabla(f, \mathcal{M})$ takes its minimal possible value $q^{-n/2}$.

For $q = 2$, Definitions 20 and 21 coincide.

A detailed study of hyperbent functions over the fields appeared in the paper by Kuz'min et al. [222]. We present here only one construction of them. The multiplicative group of the field $\mathbb{F}_{q^n}$ is the direct product of $(\mathbb{F}_{q^{n/2}}, \cdot)$ and the cyclic group V of order $q^{n/2} + 1$. Suppose that $z_{a,d}$ is equal to 1 (0) for $a, d \in \mathbb{F}_q$ whenever a and d are equal (distinct).

Theorem 129. *Take a function $g : V \to \mathbb{F}_q$ such that there is $d \in \mathbb{F}_q$ for which the number of solutions to $g(x) = a$ in V is equal to $q^{(n/2)-1} + z_{a,d}$, where $a \in \mathbb{F}_q$. Then*

$$f : \mathbb{F}_{q^n} \to \mathbb{F}_q, \quad f(0) = d, \quad f(x) = g(x^{q^{n/2}-1}) \quad \text{for } x \neq 0$$

is a hyperbent function.

For more on this topic, see [223, 388].

In 2007 and 2008 Ivanov [184, 185] also studied monomial approximations of Boolean functions. For instance, he showed [186] that the property of a bent function for it be hyperbent depends on the choice of the basis for expressing it.

In 2008 Charpin and Gong [79] studied an explicit trace representation for some classes of hyperbent functions. They presented an infinite class of monomial functions which are not hyperbent. This result indicated that Kloosterman sums on $\mathbb{F}_{2^n}$ cannot be zero at some points. Recall that Kloosterman sums naturally appear in the study of bent and hyperbent functions in trace forms. The *binary Kloosterman sums* on $\mathbb{F}_{2^n}$ are

$$K_n(a) = \sum_{x \in \mathbb{F}_{2^n}} (-1)^{\text{tr}(ax+(1/x))}, \quad \text{where } a \in \mathbb{F}_{2^n}.$$

It should be assumed that $\text{tr}(1/x) = 0$ if $x = 0$.

For functions with multiple trace terms, Charpin and Gong [79] expressed their spectra by means of Dickson polynomials. A new tool to describe these hyperbent functions was introduced.

Hyperbent functions have been intensively studied by Mesnager. In 2009 and 2011 she proposed [267, 270] a new family of hyperbent functions in polynomial form—namely, an infinite class over $\mathbb{F}_{2^n}$, $n/2$ is odd, that has the form

$$f(x) = \text{tr}_1^{o(s_1)}(ax^{s_1}) + \text{tr}_1^{o(s_2)}(bx^{s_2}),$$

where $o(s_i)$ denotes the size of the cyclotomic class modulo $2^n - 1$ containing s_i, and the coefficients are taken from the subfields $\mathbb{F}_{2^{o(s_1)}}$ and $\mathbb{F}_{2^{o(s_2)}}$. Mesnager proved that the exponents $s_1 = 3(2^{n/2} - 1)$ and $s_2 = (2^n - 1)/3$, where $a \in \mathbb{F}_{2^n}^*$ and $b \in \mathbb{F}_4$, provide a construction of hyperbent functions over $\mathbb{F}_{2^n}$. An explicit characterization of these functions in terms of Kloosterman and cubic sums was also given [267, 270].

In 2010 Mesnager [268] proposed a survey of constructions of hyperbent functions. It seems that Mesnager was not familiar with several important pieces of work in this area, such as the papers by Kuz'min et al [220, 221], otherwise, these papers should have been cited.

In 2010 Mesnager [269] continued the study initiated by Charpin and Gong [79]. A subclass of $\mathcal{PS}^-$ was investigated with respect to the property of hyperbentness. In 2011 new classes of hyperbent Boolean functions with multiple trace forms were studied by Wang et al. [370]. In 2012 Mesnager and Flori [275] studied hyperbent functions with multiple trace terms (including binomial functions) via Dillon-like exponents. They showed how the approach developed in [270] to extend the Charpin-Gong family [79], which was also used by Wang et al. to obtain another similar extension, fits in a much more general setting.

In 2012 transformations of the argument of a hyperbent function by action of the general linear group were studied by Ivanov [187]. He proved that the class of hyperbent functions is not closed under such transformations. It is the first step to study automorphism group of hyperbent functions.

In 2013 multiple-valued hyperbent functions were studied by Moraga et al. [280].

17.5 BENT FUNCTIONS OF HIGHER ORDER

This is a quite natural direction closely related to nonlinear generalizations of various methods of cryptanalysis.

It is known that the efficiency of approximating a bent function by linear functions is the lowest. When we extend the class of linear functions, it is natural to consider for approximations the Boolean functions of degree at most r, where $2 \leqslant r \leqslant n - 1$. This leads to the concept of *order r nonlinearity* $N_r(f)$ of a Boolean function f as the Hamming distance from f to all functions of this type.

Definition 22. A Boolean function at the maximal distance from all functions of degree at most r is called a *bent function of order r*.

The difficulty consists in determining this maximal possible value of $N_r(f)$. For $r \geqslant 2$, it is an open problem, better known in coding theory as determining the covering radius of the order r Reed-Muller code. Some estimates for $N_r(f)$ are known, as are its asymptotic value, connections to other cryptographic parameters, and so on. For more details on this topic, see the 2008 papers by Carlet [44],[45].

Cryptographic aspects of the higher-order bent functions were considered in 1999 by Iwata and Kurosawa [188].

With respect to bent functions of the second order, we mention the following papers. In 2011 Garg and Gangopadhyay [136] proposed a lower bound of the second-order nonlinearities of bent functions. Lower bounds of second-order nonlinearities of cubic bent functions constructed by concatenating Gold functions were studied in 2011 by Gode and Gangopadhyay, [141]. Also in 2011, Kolokotronis and Limniotis [204] considered Maiorana-McFarland functions with high second-order nonlinearity. Second-order nonlinearities of some special bent functions were considered by Gangopadhyay and Singh [133] in 2012 and by Tang et al. [350] in 2013.

17.6 k-BENT FUNCTIONS

In 2007, the author [354] introduced the following concept whose main idea is to consider approximating functions distinct from linear functions, but analogous to them in some sense.

Take binary vectors x and y of length n and an arbitrary integer k satisfying $1 \leqslant k \leqslant n/2$. Define the binary operation

$$\langle x, y \rangle_k = \langle x, y \rangle \oplus \left(\bigoplus_{i=1}^{k} \bigoplus_{j=i}^{k} (x_{2i-1} \oplus x_{2i})(x_{2j-1} \oplus x_{2j})(y_{2i-1} \oplus y_{2i}) \right.$$

$$\left. (y_{2j-1} \oplus y_{2j}) \right),$$

which serves as a nonlinear analogue of the inner product. Observe that in this operation the components of the vectors are inequivalent in the following sense: the first $2k$ components of each of them appear in both quadratic and linear terms, while the rest, appear only in the linear terms.

The function

$$W_f^{(k)}(y) = \sum_{x \in \mathbb{F}_2^n} (-1)^{\langle x, y \rangle_k \oplus f(x)}$$

is called the k-*Walsh-Hadamard transform* of a Boolean function f. For $k = 1$, we have an expression equivalent to the ordinary Walsh-Hadamard transform. It is easy to see that Parseval's equality

$$\sum_{y \in \mathbb{F}_2^n} \left(W_f^{(k)}(y) \right)^2 = 2^{2n}$$

holds. A function f is called a *k-bent function with a fixed order of variables* if all its coefficients $W_f^{(j)}(y)$, $j = 1, \ldots, k$, are equal to $\pm 2^{n/2}$. These functions are considered in [354, 356].

Let us consider a more general definition, which is free from the restrictions on the order of variables.

Definition 23. A Boolean function f in n variables is called a *k-bent function* if $W_{f \circ \pi}^{(j)}(y) = \pm 2^{n/2}$ for an arbitrary permutation $\pi \in S_n$, every $j = 1, \ldots, k$, and every vector y.

Let us explain this definition. Consider the set of functions

$$A_n^k(\pi) = \left\{ \langle \pi(x), y \rangle_k \oplus a \mid y \in \mathbb{F}_2^n, \ a \in \mathbb{F}_2 \right\}$$

in n variables. The vectors of values of the functions of every class $A_n^k(\pi)$ constitute a binary Hadamard code. This code is nonlinear (for $k > 1$), but in the space $\mathbb{Z}_4^n$ there is a linear preimage of it under a simple mapping; see [354, 356] for more details. The approach was based on $\mathbb{Z}_4$-linear Hadamard codes described by Krotov [215], [216] in 2000–2001 (for general information on $\mathbb{Z}_4$-linear codes see [150]). Thus, we may regard the functions in $A_n^k(\pi)$ as analogues of affine functions. Observe that they are quadratic. The *k-nonlinearity* of a Boolean function f is the minimal Hamming distance $N_f^{(k)}$ from it to the set of all functions of the form $\langle \pi(x), y \rangle_k \oplus a$, where π is an arbitrary permutation. We have

$$N_f^{(k)} = 2^{n-1} - \frac{1}{2} \max_{\pi \in S_n} \max_{y \in \mathbb{F}_2^n} \left| W_{f \circ \pi}^{(k)}(y) \right|.$$

Therefore, a k-bent function is a function with maximal $N_f^{(j)}$—that is, $N_f^{(j)} = 2^{n-1} - 2^{(n/2)-1}$, for all $j = 1, \ldots, k$. Thus, it is simultaneously maximally distant from all classes of functions $A_n^j(\pi)$, $\pi \in S_n$ and $j = 1, \ldots, k$.

Observe that 1-bent functions coincide with ordinary bent functions. As k increases, the nonlinear properties of functions become stronger. Thus, the most interesting problem apparently is to describe the class of all $(n/2)$-bent functions. As implied in [354], this class is nonempty. For every even n, it contains, for instance, all symmetric bent functions:

$$f(x) = \bigoplus_{i=1}^{n} \bigoplus_{j=i+1}^{n} x_i x_j,$$

and $f(x) \oplus 1, f(x) \oplus \sum_{i=1}^{n} x_i, f(x) \oplus \sum_{i=1}^{n} x_i \oplus 1$, which were characterized by Savicky [321] in 1994.

For $n = 4$, all $(n/2)$-bent functions are described in [358]. There are 128 quadratic functions with the quadratic part of one of the four types

$$x_1 x_2 \oplus x_3 x_4, \quad x_1 x_3 \oplus x_2 x_4, \quad x_1 x_4 \oplus x_2 x_3,$$

$$x_1 x_2 \oplus x_1 x_3 \oplus x_1 x_4 \oplus x_2 x_3 \oplus x_2 x_4 \oplus x_3 x_4,$$

and an arbitrary linear part. See also [355, 357] on this topic.

REFERENCES

[1] Adams C. On immunity against Biham and Shamir's "differential cryptanalysis". Inf Process Lett 1992;41:77-80.

[2] Adams C. Constructing symmetric ciphers using the CAST design procedure. Des Codes Crypt 1997;12(3):283-316.

[3] Adams C, Tavares S. The structured design of cryptographically good S-boxes. J Cryptol 1990;3(1):27-43.

[4] Adams C, Tavares S. Generating and counting binary bent sequences. IEEE Trans Inform Theory 1990;36:1170-3.

[5] Ageev DV. Foundations of the theory of linear selection. A code channel division. In: Collection of scientific papers of the Leningrad experimental institute of communications; 1935 [in Russian].

[6] Agibalov GP. Selected theorems of the initial course of cryptography. Tomsk: Tomsk State University; 2005. 116p. [in Russian].

[7] Agievich SV. On the representation of bent functions by bent rectangles. In: Proceedings of the fifth international Petrozavodsk conference on probabilistic methods in discrete mathematics (Petrozavodsk, Russia, June 1-6, 2000). Boston: VSP; 2000. p. 121-35. Available at: URL: http://arxiv.org/abs/math/0502087.

[8] Agievich SV. On the affine classification of cubic bent functions, Cryptology ePrint Archive, report 2005/044; 2005. Available at: URL: http://eprint.iacr.org/.

[9] Agievich SV. Bent rectangles. In: Proceedings of the NATO Advanced Study Institute on Boolean functions in cryptology and information security (Zvenigorod, Russia, September 8-18, 2007). Netherlands: IOS Press; 2008. p. 3-22. Available at: URL: http://arxiv.org/abs/0804.0209.

[10] Agievich SV. Private communication; 2014.

[11] Agievich S, Gorodilova A, Kolomeec N, Nikova S, Preneel B, Rijmen V, et al. Problems, solutions and experience of the first international student's Olympiad in cryptography. Prikl Diskr Matem 2015; In print. See also on Cryptology ePrint Archive, Report 2015/534 (the concrete number of the report will appear during the week), available at, URL: http://eprint.iacr.org/.

[12] Akyildiz E, Guloglu IS, Ikeda M. A note of generalized bent functions. J Pure Appl Algebra 1996;106(1):1 9.

[13] Ambrosimov AS. Properties of bent functions of q-valued logic over finite fields. Discret Math Appl 1994;4(4):341-50.

[14] Baignères T, Junod P, Vaudenay S. How far can we go beyond linear cryptanalysis? In: Advances in cryptology—ASIACRYPT'04, proceeding of the 10th international conference on the theory and applications of cryptology and information security (Jeju Island, Korea, December 5-9, 2004). Lecture notes in computer science, vol. 3329. Berlin: Springer; 2004. p. 432-50.

[15] Bajric S, Pasalic E, Ribic-Muratovic A, Gangopadhyay S. On generalized bent functions with Dillon's exponents. Inf Process Lett 2014;114(4):222-7.

[16] Bernasconi A, Codenotti B. Spectral analysis of Boolean functions as a graph eigenvalue problem. IEEE Trans Comput 1999;48(3):345-51.

[17] Bernasconi A, Codenotti B, VanderKam JM. A characterization of bent functions in terms of strongly regular graphs. IEEE Trans Comput 2001;50(9):984-5.

[18] Bey Ch, Kyureghyan G. An association scheme of a family of cubic bent functions. In: Proceedings of the international workshop on coding and cryptography (Versailles, France, April 16-20); 2007. p. 13-9.

[19] Bey Ch, Kyureghyan G. On Boolean functions with the sum of every two of them being bent. Des Codes Crypt 2008;49(1-3):341-6.

[20] Biham E, Dunkelman O, Keller N. Differential-linear cryptanalysis of serpent. In: Fast software encryption—FSE'2003, proceedings of the 10th international workshop (Lund, Sweden, February 24-26, 2003). Lecture notes in computer science, vol. 2887. Berlin: Springer; 2003. p. 9-21.

[21] Biham E, Shamir A. Differential cryptanalysis of DES-like cryptosystems. J Cryptol 1991;4(1):3-72.

[22] Biryukov A, De Canniere C, Quisquater M. On multiple linear approximations. In: Advances in cryptology—CRYPTO 2004, proceedings of the 24th annual international cryptology conference (Santa Barbara, California, USA, August 15-19, 2004). Lecture notes in computer science, vol. 3152. Berlin: Springer-Verlag; 2004. p. 1-22.

[23] Borst J, Preneel B, Vandewalle J. Linear cryptanalysis of RC5 and RC6. In: Fast software encryption—FSE'99, proceedings of the 6th international workshop (Rome, Italy, March 24-26, 1999). Lecture notes in computer science, vol. 1636. Berlin: Springer; 1999. p. 16-30.

[24] Braeken A. Cryptographic properties of Boolean functions and S-boxes. Ph.d. thesis; 2006. Available at: URL: http://www.cosic.esat.kuleuven.be/publications/thesis-129.pdf.

[25] Budaghyan L. Construction and analysis of cryptographic functions. Habilitation thesis; 2013.

[26] Budaghyan L, Carlet C, Helleseth T, Kholosha A, Mesnager S. Further results on niho bent functions. IEEE Trans Inform Theory 2012;58(11):6979-85.

[27] Budaghyan L, Carlet C, Helleseth T, Kholosha A. Generalized bent functions and their relation to Maiorana-Mcfarland class. In: Proceedings of ISIT—IEEE international symposium on information theory (Cambridge, USA, July 1-6, 2012); 2012.

[28] Buryakov ML, Logachev OA. On the level of affinity of Boolean functions. Discret Math Appl 2005;15(5):479-88.

[29] Buttyan L, Vajda I. Searching for the best linear approximation of des-like cryptosystems. Electron Lett 1995;31(11):873-4.

[30] Canteaut A. Analysis and design of symmetric ciphers. Habilitation thesis; 2006.

[31] Canteaut A, Charpin P. Decomposing bent functions. IEEE Trans Inform Theory 2003;49:2004-19.

[32] Canteaut A, Charpin P, Kuyreghyan G. A new class of monomial bent functions. Finite Fields Appl 2008;14(1):221-41.

[33] Canteaut A, Daum M, Dobbertin H, Leander G. Finding nonnormal bent functions. Discret Appl Math 2006;154(2):202-18.

[34] Canteaut A, Videau M. Higher order differential attacks on iterated block ciphers using almost bent round functions. In: Proceedings of ISIT—IEEE international symposium on information theory (Lausanne, Switzerland, June 30-July 5, 2002); 2002. p. 209.

[35] Carlet C. Partially-bent functions. Des Codes Crypt 1993;3(2):135-45.

[36] Carlet C. Two new classes of bent functions. In: Advances in cryptography—EUROCRYPT'93, workshop on the theory and application of cryptographic techniques (Lofthus, Norway, May 23-27, 1993). Lecture notes in computer science, vol. 765. Berlin: Springer; 1993. p. 77-101.

[37] Carlet C. Generalized partial spreads. IEEE Trans Inform Theory 1995;41(5):1482-7.

[38] Carlet C. A construction of bent functions. In: Finite fields and applications. Lecture series, vol. 233, London Mathematical Society; 1996. p. 47-58.

[39] Carlet C. Hyper-bent functions. In: International conference on the theory and applications of cryptology—PRAGOCRYPT'96. Prague: Czech Technical University Publishing House; 1996. p. 149-55.

[40] Carlet C. On cryptographic complexity of Boolean functions. In: Mullen GL, Stichtenoth H, Tapia-Recillas H, editors. Proceedings of the sixth conference on finite fields with applications to coding theory, cryptography and related areas. Berlin: Springer; 2002. p. 53-69.

[41] Carlet C. On the confusion and diffusion properties of Maiorana—McFarland's and extended Maiorana—McFarland's functions. J Complexity 2004;20:182-204. Special issue "Complexity issues in coding theory and cryptography" dedicated to Prof. Harald Niederreiter on the occasion of his 60th birthday.

[42] Carlet C. Designing bent functions and resilient functions from known ones, without extending their number of variables. In: Proceedings of ISIT—IEEE international symposium on information theory (Adelaide, Australia, September 4-9, 2005); 2005. p. 1096-100.

[43] Carlet C. On bent and highly nonlinear balanced/resilient functions and their algebraic immunities. In: Applied algebra, Algebraic algorithms and error correcting codes, Las Vegas, USA; 2006. p. 1-28.

[44] Carlet C. Recursive lower bounds on the nonlinearity profile of boolean functions and their applications. IEEE Trans Inform Theory 2008;54(3):1262-72.

[45] Carlet C. On the higher order nonlinearities of Boolean functions and S-boxes, and their generalizations. In: Proceedings of the fifth international conference on sequences and their applications—SETA'2008 (Lexington, Kentucky, USA, September 14-18, 2008). Lecture notes in computer science, vol. 5203. Berlin: Springer; 2008. p. 345-67.

[46] Carlet C. Boolean functions for cryptography and error-correcting codes. In: Hammer P, Crama Y, editors. Boolean models and methods in mathematics, computer science, and engineering, chap. 8. UK: Cambridge University Press; 2010. p. 257-397. URL: www.math.univ-paris13.fr/~carlet/.

[47] Carlet C. Vectorial Boolean functions for cryptography. In: Hammer P, Crama Y, editors. Boolean models and methods in mathematics, computer science, and engineering, chap. 9. UK: Cambridge University Press; 2010. p. 398-472. URL: www.math.univ-paris13.fr/~carlet/.

[48] Carlet C. Relating three nonlinearity parameters of vectorial functions and building APN functions from bent functions. Des Codes Crypt 2011;59(1-3):89-109.

[49] Carlet C. Open problems on binary bent functions. In: Proceedings of the conference "Open problems in mathematical and computational sciences" (Istanbul, Turkey, September 18-20); 2013

[50] Carlet C. Open questions on nonlinearity and on APN functions. In: Proceedings of WAIFI (Gebze, Turkey, September 27-28, 2014); 2014. p. 83-107.

[51] Carlet C, Charpin P, Zinoviev V. Codes, bent functions and permutations suitable for DES-like cryptosystems. Des Codes Crypt 1998;15(2):125-56.

[52] Carlet C, Danielsen LE, Parker MG, Solé P. Self dual bent functions. In: Proceedings of the fourth international conference, BFCA—Boolean functions: cryptography and applications (Copenhagen, Denmark, May 19–21, 2008); 1998. p. 39-52. Full version in Int J Inform Coding Theory 2010;1:384-399.

[53] Carlet C, Ding C. Highly nonlinear mappings. J Complexity 2004;20(2-3):205-44.

[54] Carlet C, Ding C. Nonlinearities of S-boxes. Finite Fields Appl 2007;13(1):121-35.

[55] Carlet C, Ding C, Niederreiter H. Authentication schemes from highly nonlinear functions. Des Codes Crypt 2006;40(1):71-9.

[56] Carlet C, Dobbertin H, Leander G. Normal extensions of bent functions. IEEE Trans Inform Theory 2004;50(11):2880-5.

[57] Carlet C, Dubuc S. On generalized bent and q-ary perfect nonlinear functions. Finite Fields Appl 2001:81-94.

[58] Carlet C, Gaborit P. Hyper-bent functions and cyclic codes. J Combin Theory Ser A 2006;113(3):466-82.

[59] Carlet C, Gao G, Liu W. A secondary construction and a transformation on rotation symmetric functions, and their action on bent and semi-bent functions. J Combin Theory Ser A 2014;127:161-75.

[60] Carlet C, Guillot P. A characterization of binary bent functions. J Combin Theory Ser A 1996;76(2):328-35.

[61] Carlet C, Guillot P. An alternate characterization of the bentness of binary functions, with uniqueness. Des Codes Crypt 1998;14:133-40.

[62] Carlet C, Helleseth T, Kholosha A, Mesnager S. On the dual of bent functions with 2^r Niho exponents. In: Proceedings of the ISIT—IEEE international symposium on information theory (St Petersburg, Russia, July 31-August 5, 2011); 2011. p. 703-07.

[63] Carlet C, Klapper A. Upper bounds on the numbers of resilient functions and of bent functions. In: Proceedings of the 23rd symposium on information theory (Benelux, Belgium, May, 2002); 2002. p. 307-14. The full version will appear in Lecture Notes dedicated to Philippe Delsarte. Available at: URL: http://www.cs.engr.uky.edu/~klapper/ps/bent.ps.

[64] Carlet C, Ku-Cauich C, Tapia-Recillas H. Bent functions on a galois ring and systematic authentication codes. Adv Math Commun 2012;6(2):249-58.

[65] Carlet C, Mesnager S. On Dillon's class H of bent functions, Niho bent functions and o-polynomials. J Combin Theory Ser A 2011;118(8):2392-410.

[66] Carlet C, Mesnager S. On semibent Boolean functions. IEEE Trans Inform Theory 2012;58(5):3287-92.

[67] Carlet C, Prouff E. On plateaued functions and their constructions. In: Fast software encryption—FSE'2003 (Proceedings of the 10th international workshop, Lund, Sweden, February 24-26, 2003). Lecture notes in computer science, vol. 2887. Berlin: Springer; 2003. p. 54-73.

[68] Carlet C, Yucas J. Piecewise constructions of bent and almost optimal Boolean functions. Des Codes Crypt 2005;37(3):449-64.

[69] Carlet C, Zhang F, Hu Yu. Secondary constructions of bent functions and their enforcement. Adv Math Commun 2012;6(3):305-14.

[70] CAST-128. Rfc 2144—the cast-128 encryption algorithm; 1997. URL: http://www.faqs.org/rfcs/rfc2144.html.

[71] Çeşmelioğlu A, McGuire G, Meidl W. A construction of weakly and non-weakly regular bent functions. J Combin Theory Ser A 2012;119(2):420-9.

[72] Çeşmelioğlu A, Meidl W. Bent functions of maximal degree. IEEE Trans Inform Theory 2012;58(2):1186-90.

[73] Çeşmelioğlu A, Meidl W, Pott A. On the dual of (non)-weakly regular bent functions and self-dual bent functions. Adv Math Commun 2013;7(4):425-40.

[74] Çeşmelioğlu A, Meidl W, Pott A. Generalized Maiorana-McFarland class and normality of p-ary bent functions. Finite Fields Appl 2013;24:105-17.

[75] Çeşmelioğlu A, Meidl W. A construction of bent functions from plateaued functions. Des Codes Crypt 2013;66(1-3):231-42.

[76] Chabaud F, Vaudenay S. Links between differential and linear cryptanalysis. In: Advances in cryptology—EUROCRYPT '94, proceedings of the international conference on the theory and application of cryptographic techniques (Perugia, Italy, May 9-12, 1994). Lecture notes in computer science, vol. 950. Berlin: Springer; 1995. p. 356-65.

[77] Charnes C, Rotteler M, Beth T. Homogeneous bent functions, invariants, and designs. Des Codes Crypt 2002;26(1-3):139-54.

[78] Charpin P. Normal Boolean functions. J Complexity 2004;20:245-65.

[79] Charpin P, Gong G. Hyperbent functions, Kloosterman sums, and Dickson polynomials. IEEE Trans Inform Theory 2008;54(9):4230-8.

[80] Charpin P, Kyureghyan G. Cubic monomial bent functions: a subclass of M. SIAM J Discret Math 2008;22(2):650-65.

[81] Charpin P, Pasalic E, Tavernier C. On bent and semi-bent quadratic Boolean functions. IEEE Trans Inform Theory 2005;51(12):4286-98.

[82] Chase PJ, Dillon JF, Lerche KD. Bent functions and difference sets. NSA R41 Technical Paper; September, 1970.

[83] Chee S, Lee S, Kim K. Semi-bent functions. In: Advances in cryptology—ASIACRYPT'94, proceedings of the 4th international conference on the theory and applications of cryptology (Wollongong, Australia, November 28-December 1, 1994). Lecture notes in computer science, vol. 917. Berlin: Springer; 1995. p. 107-18.

[84] Chee YM, Tan Y, Zhang X. Strongly regular graphs constructed from p-ary bent functions. J Algebr Combinator 2011;34(2):251-66.

[85] Chen H, Cao X. Some semi-bent functions with polynomial trace form. J Syst Sci Complexity 2014;27(4):777-84.

[86] Chen YQ, Polhill J. Partial difference sets and amorphic group schemes from pseudo-quadratic bent functions. J Algebr Combinator 2013;37(1):11-26.

[87] Cheremushkin AV. Methods of affine and linear classifications of Boolean functions. Trudy Diskret Matem (Bull Discret Math) M: Fizmatlit 2001;4:273-314 [in Russian].

[88] Chung H, Kumar PV. A new general construction for generalized bent functions. IEEE Trans Inform Theory 1989;35(1):206-9.

[89] Clark JA, Jacob JL. Two-stage optimisation in the design of Boolean functions. In: Proceedings of the 5th Australian conference on information security and privacy (Brisbane, Australia, July 10-12, 2000). Lecture notes in computer science, vol. 1841. Berlin: Springer-Verlag; 2000. p. 242-54.

[90] Clark JA, Jacob JL, Maitra S, Stănică P. Almost Boolean functions: the design of Boolean functions by spectral inversion. Comput Intell Spec Issue Evol Comput Cryptography Secur 2004;20(3):450-62.

[91] Climent JJ, Garcia FJ, Requena V. On the construction of bent functions of $n + 2$ variables from bent functions of n variables. Adv Math Commun 2008;2(4):421-31.

[92] Climent JJ, Garcia FJ, Requena V. Construction of bent functions of $2k$ variables from a basis of $\mathbb{F}_2^{2k}$. Int J Comput Math 2012;89(7):863-80.

[93] Climent JJ, Garcia FJ, Requena V. A construction of bent functions of $n + 2$ variables from a bent function of n variables and its cyclic shifts. Algebra 2014;2014:11. Article ID 701298.

[94] Cornell University, Department of Mathematics. Annual report 2001-2002. URL: www.math.cornell.edu/News/AnnRep/AR01-02.pdf.

[95] Cusick TW, Stănică P. Fast evaluation, weights and nonlinearity of rotation-symmetric functions. Discret Math 2002;258:289-301.

[96] Cusick TW, Stănică P. Cryptographic Boolean functions and applications. USA: Acad. Press. Elsevier; 2009. 245p.

[97] Cusick TW, Li Y, Stănică P. On a combinatoric conjecture. Cryptology ePrint Archive, report 2009/554; 2009. Available at: URL: http://eprint.iacr.org/.

[98] Daemen J, Govaerts R, Vandevalle J. Correlation matrices. In: Fast software encryption—FSE'95, proceedings of the second international workshop (Leuven, Belgium, December 14-16, 1994). Lecture notes in computer science, vol. 1008. Berlin: Springer; 1995. p. 275-85.

[99] Dalai DK, Maitra S, Sarkar S. Results on rotation symmetric bent functions. Discret Math 2009;309(8):2398-409. Proceedings of the 2nd international workshop on Boolean functions: cryptography and applications—BFCA, 2006 (Rouen, France, March 13-15, 2006).

[100] Danielsen LE, Parker M, Solé P. The Rayleigh quotient of bent functions. In: Proceedings of the 12th IMA conference on cryptography and coding (Cirencester, England, December 15-17, 2009), vol. 5921; 2009. p. 418-32.

[101] Daum M, Dobbertin H, Leander G. An algorithm for checking normality of Boolean functions. Discret Appl Math 2003:133-43.

[102] Delsarte P. An algebraic approach to the association schemes of coding theory. Ph.d. thesis; 1973.

[103] Dempwolff U. Automorphisms and equivalence of bent functions and of difference sets in elementary Abelian 2-groups. Commun Algebra 2006;34(3):1077-131.

[104] Dempwolff U. Homepage of U. Dempwolff. see the section "Bent Functions in Dimensions 8,10,12"; 2009. URL: http://www.mathematik.uni-kl.de/~dempw/.

[105] Deng G. A note on a combinatorial conjecture. Open J Discret Math 2013;3:49-52.

[106] Detombe J, Tavares S. Constructing large cryptographically strong S-boxes. In: Advances in cryptology—AUSCRYPT'92 (Gold Coast, Queensland, Australia, December 13-16, 1992). Lecture notes in computer science, vol. 718. Berlin: Springer; 1993. p. 165-81.

[107] Dillon JF. A survey of bent functions. NSA Tech J 1972; (special issue):191-215.

[108] Dillon JF. Elementary Hadamard difference sets. Ph.d. thesis; 1974.

[109] Dillon JF, Dobbertin H. New cyclic difference sets with Singer parameters. Finite Fields Appl 2004;10(3):342-89.

[110] Dillon JF, McGuire G. Near bent functions on a hyperplane. Finite Fields Appl 2008;14:715-20.

[111] Dobbertin H. Construction of bent functions and balanced Boolean functions with high nonlinearity. In: Fast software encryption—FSE'95, second international workshop (Leuven, Belgium, December 14-16, 1994). Lecture notes in computer science, vol. 1008. Berlin: Springer; 1995. p. 61-74.

[112] Dobbertin H, Leander G. A survey of some recent results on bent functions. In: Sequences and their applications—SETA 2004, third international conference (Seul, Korea, October 24-28, 2004), revised selected papers. Lecture notes in computer science, vol. 3486. Berlin: Springer; 2005. p. 1-29.

[113] Dobbertin H, Leander G. Cryptographer's toolkit for construction of 8-bit bent functions. Cryptology ePrint Archive, report 2005/089; 2005. Available at: URL: http://eprint.iacr.org/.

[114] Dobbertin H, Leander G. Bent functions embedded into the recursive framework of F-bent functions. Des Codes Crypt 2008;49(1-3):3-22.

[115] Dobbertin H, Leander G, Canteaut A, Carlet C, Felke P, Gaborit Ph. Construction of bent functions via Niho power functions. J Combin Theory Ser A 2006;113(5):779-98.

[116] Dong D, Qu L, Fu S, Li C. New constructions of semi-bent functions in polynomial forms. Math Comput Model 2013;57(5-6):1139-47.

[117] Feng KQ. Generalized bent functions and class group of imaginary quadratic fields. Sci China Ser A: Math Phys Astron 2001;44(5):562-70.

[118] Feng KQ, Liu FM. New results on the nonexistence of generalized bent functions. IEEE Trans Inform Theory 2003;49(11):3066-71.

[119] Feulner T, Sok L, Solé P, Wassermann A. Towards the classification of self-dual bent functions in eight variables. Des Codes Crypt 2013;68(1-3):395-406.

[120] Filiol E, Fountain C. Highly nonlinear balanced Boolean functions with a good correlation immunity. In: Advances in cryptography—EUROCRYPT'98, workshop on the theory and application of cryptographic techniques. Berlin: Springer; 1998. p. 475-88.

[121] Filyuzin SY. On algebraic immunity of bent functions from the Dillon's class. Diskretn Anal Issled Oper (Discret Anal Oper Res) 2014;21(5):67-75 [in Russian]. English translation appears soon in J Appl Ind Math.

[122] Filyuzin SY. Algebraic immunity of Dillon's bent functions constructed with monomial functions. Prikl Diskr Matem (Discrete Appl Math) in press; 8(Suppl.) (in Russian) 2015.

[123] Frolova AA. The essential dependence of Kasami bent functions on the products of variables. Diskretn Anal Issled Oper (Discret Anal Oper Res) 2013;20(1):72-99 [in Russian]. English translation appears soon in J Appl Ind Math.

[124] Fuller JE. Analysis of affine equivalent Boolean functions for cryptography. Ph.d. thesis. Brisbane, Australia; 2003. Available at: URL: http://eprints.qut.edu.au/15828/.

[125] Fuller JE, Dawson E, Millan W. Evolutionary generation of bent functions for cryptography. In: Proceedings of the 2003 congress on evolutionary computation, vol. 3; 2003. p. 1655-61.

[126] Gabidulin EM. Partial classification of sequences with perfect auto-correlation and bent functions. In: Proceedings of the IEEE international symposium on information theory (Whistler, Canada, September 17-22, 1995); 1995. p. 467.

[127] Gangopadhyay S. Affine inequivalence of cubic Maiorana-McFarland type bent functions. Discret Appl Math 2013;161(7-8):1141-6.

[128] Gangopadhyay S, Joshi A, Leander G, Sharma RK. A new construction of bent functions based on $\mathbb{Z}$-bent functions. Des Codes Crypt 2013;66(1-3): 243-56.

[129] Gangopadhyay S, Pasalic E, Singh BK. On upper bounds on algebraic immunity of some PSap and Niho bent functions. In: Proceedings of the 8th international ICST conference on communications and networking in China—CHINACOM (Guilin, China, August 14-16, 2013); 2013. p. 379-83.

[130] Gangopadhyay S, Pasalic E, Stanica P. A note on generalized bent criteria for Boolean functions. IEEE Trans Inform Theory 2013;59(5):3233-6.

[131] Gangopadhyay S, Sharma D. On construction of non-normal Boolean functions. Australas J Combin 2007;38:267-72.

[132] Gangopadhyay S, Sharma D, Sarkar S, Maitra S. On affine (non) equivalence of bent functions. In: CECC'08—Central European conference on cryptography (Graz, Austria, July 2-4, 2008); 2008. Available at: URL: www.math.tugraz.at/~cecc08/abstracts/cecc08_abstract_25.pdf.

[133] Gangopadhyay S, Singh BK. On second-order nonlinearities of some D-0 type bent functions. Fund Inform 2012;114(3-4):271-85.

[134] Gao G, Zhang X, Liu W, Carlet C. Constructions of quadratic and cubic rotation symmetric bent functions. IEEE Trans Inform Theory 2012;58(7): 4908-13.

[135] Garcia FJ, Requena V, Tomas V. Generating pseudo-random sequences from cellular automata and bent functions. In: Proceedings of the 6th WSEAS international conference on informational security and privacy (Canary Islands, Spain, December 14-16, 2007); 2007. p. 40-3.

[136] Garg M, Gangopadhyay S. A lower bound of the second-order nonlinearities of Boolean bent functions. Fund Inform 2011;111(4):413-22.

[137] Glukhov MM. Planar mappings of finite fields and their generalizations. In: Presentation for the conference "Algebra and logic: theory and applications", (Krasnoyarsk, Russia, July 21-27); 2013 [in Russian].

[138] Glukhov MM. On perfect and almost perfect nonlinear functions. In: Presentation for the Russian symposium on applied and industrial mathematics (Sochi—Dagomys, Russia, September 28-October 5); 2014 [in Russian].

[139] Glukhov MM, Zakrevskii PN. On additivity and affinity coefficients of discrete functions. Discret Math Appl 2012;22(1):1-22.

[140] Gmurman VE. Probability theory and mathematical statistics. Moscow: Higher Educ.; 2006.

[141] Gode R, Gangopadhyay S. On lower bounds of second-order nonlinearities of cubic bent functions constructed by concatenating gold functions. Int J Comput Math 2011;88(15):3125-35.

[142] Gong G. Correlation of multiple bent function signal sets. In: Proceedings of the IEEE information theory workshop information theory for wireless networks (Solstrand, Norway, July 1-6, 2007); 2007. p. 148-52.

[143] Gong G, Golomb SW. Transform domain analysis of DES. IEEE Trans Inform Theory 1999;45(6):2065-73.

[144] Gong G, Helleseth T, Hu H, Kholosha A. On the dual of certain ternary weakly regular bent functions. IEEE Trans Inform Theory 2012;58(4):2237-43.

[145] Grocholewska-Czuryło A. A study of differences between bent functions constructed using Rothaus method and randomly generated bent functions. J Telecommun Inform Technol 2003;4:19-24. Available at: URL: www.itl.waw.pl/czasopisma/JTIT/2003/4/19.pdf.

[146] Grocholewska-Czuryło A, Stoklosa J. Generating bent functions. In: Proceedings of the 8th international conference on advanced computer systems (Mielno, Poland, October 17-19, 2001). Springer international series in engineering and computer science, vol. 664; 2002. p. 361-70.

[147] Guillot P. Completed GPS covers all bent functions. J Combin Theory Ser A 2001;93(2):242-60.

[148] Gupta KC, Nawaz Y, Gong G. Upper bound for algebraic immunity on a subclass of Maiorana-McFarland class of bent functions. Inf Process Lett 2011;111(5):247-9.

[149] Hall MJ. Combinatorial theory. London: Blaisdell Publishing; 1967. 424p.

[150] Hammons AR, Kumar PV, Calderbank AR, Sloane NJA, Solé P. The $\mathbb{Z}_4$-linearity of Kerdock, Preparata, Goethals, and related codes. IEEE Trans Inform Theory 1994;40(2):301-19.

[151] Harpers C, Kramer GG, Massey JL. A generalization of linear cryptanalysis and the applicability of Matsui's piling-up lemma. In: Advances in cryptology—EUROCRYPT'95—international conference on the theory and application of cryptographic techniques (Saint-Malo, France, May 21-25, 1995). Lecture notes in computer science, vol. 921. Berlin: Springer; 1995. p. 24-38.

[152] Hawkes P, O'Connor L. On applying linear cryptanalysis to IDEA. In: Advances in cryptology—ASIACRYPT'96—international conference on the theory and applications of cryptology and information security (Kyongju, Korea, November 3-7, 1996). Lecture notes in computer science, vol. 1163. Berlin: Springer; 1996. p. 105-15.

[153] He Y, Ma W, Kang P. On semi-bent functions with Niho exponents. Sci China Inform Sci 2012;55(7):1624-30.

[154] He Y, Ma W. Balanced semi-bent functions with high algebraic degrees. IEICE Trans Fundam Electron Commun Comput Sci 2011;E94.A(3):1019-22.

[155] Hell M, Johansson T, Meier W. A stream cipher proposal: Grain-128. eSTREAM ECRYPT Stream Cipher Project; 2006. URL: http://www.ecrypt.eu.org/stream/grainpf.html.

[156] Helleseth T, Hollmann HDL, Kholosha A, Wang Z, Xiang Q. Proofs of two conjectures on ternary weakly regular bent functions. IEEE Trans Inform Theory 2009;55(11):5272-83.

[157] Helleseth T, Kholosha A. Monomial and quadratic bent functions over the finite fields of odd characteristic. IEEE Trans Inform Theory 2006;52(5):2018-32. Proceedings of the IEEE information theory workshop on coding and complexity (Rotorua, New Zeland, August-September, 2005).

[158] Helleseth T, Kholosha A. On the dual of monomial quadratic p-ary bent functions. In: Proceedings of the international workshop on sequences, subsequences and consequences (Los Angeles, USA, May 31-June 2, 2007). Lecture notes in computer science, vol. 4893. Berlin: Springer; 2007. p. 50-61.

[159] Helleseth T, Kholosha A. Sequences, bent functions and Jacobsthal sums. In: Proceedings of the 6th international conference on sequences and their applications (Paris, France, September 13-17, 2010). Lecture notes in computer science, vol. 6338. Berlin: Springer; 2010. p. 416-29.

[160] Helleseth T, Kholosha A. New binomial bent functions over the finite fields of odd characteristic. IEEE Trans Inform Theory 2010;56(9):4646-52.

[161] Helleseth T, Kholosha A. Crosscorrelation of m-sequences, exponential sums, bent functions and jacobsthal sums. Cryptogr Commun 2011;3(4):281-91.

[162] Helleseth T, Kholosha A. Bent functions and their connections to combinatorics. Surv Combinator 2013:91-126.

[163] Helleseth T, Kholosha A, Mesnager S. Bent functions and Subiaco hyperovals. In: Proceedings of the 10th international conference on finite fields and their applications (Ghent, Belgium, July 11-15, 2011). Contemporary mathematics, vol. 579; 2012. p. 91-101.

[164] Heys HM, Tavares SE. Substitution-permutation networks resistant to differential and linear cryptanalysis. J Cryptol 1996;9(1):1-19.

[165] Hodzic S, Pasalic E. Generalized bent functions—some general construction methods and related necessary and sufficient conditions. Cryptogr Commun 2015; doi:10.1007/s12095-015-0126-9.

[166] Hollmann H, Xiang Q. On binary cyclic codes with few weights. In: Proceedings of the fifth conference on finite fields and their applications (Augsburg, Germany). Berlin: Springer; 1999. p. 251-75.

[167] Hou XD. Cubic bent functions. Discret Math 1998;189:149-61.

[168] Hou XD. q-ary bent functions constructed from chain rings. Finite Fields Appl 1998;4(1):55-61.

[169] Hou XD. Bent functions, partial difference sets, and quasi-frobenius local rings. Des Codes Crypt 2000;20(3):251-68.

[170] Hou XD. On the coefficients of binary bent functions. Proc Am Math Soc 2000;128(4):987-96.

[171] Hou XD. p-ary and q-ary versions of certain results about bent functions and resilient functions. Finite Fields Appl 2004;10(4):566-82.

[172] Hou XD. On the dual of a Coulter-Matthews bent function. Finite Fields Appl 2008;14(2):505-14.

[173] Hou XD. Classification of self dual quadratic bent functions. Des Codes Crypt 2012;63(2):183-98.

[174] Hou XD. Classification of p-ary self dual quadratic bent functions, p odd. J Algebra 2013;391:62-81.

[175] Hou XD, Langevin P. Results on bent functions. J Combin Theory Ser A 1997;80:232-46.

[176] Hu H, Feng D. On quadratic bent functions in polynomial forms. IEEE Trans Inform Theory 2007;53(7):2610-15.

[177] Huang TY, You KH. Strongly regular graphs associated with bent functions. In: Proceedings of the 7th international symposium on parallel architectures, algorithms and networks (Hong Kong, China, May 10-12, 2004); 2004. p. 380-3.

[178] Hunt FH, Smith DH. The construction of orthogonal variable spreading factor codes from semi-bent functions. IEEE Trans Wirel Commun 2012;11(8): 2970-5.

[179] Hunt FH, Smith DH. The assignment of CDMA spreading codes constructed from Hadamard matrices and almost bent functions. Wirel Pers Commun 2013;72(4): 2215-27.

[180] Hyun JY, Lee H, Lee Y. Macwilliams duality and a Gleason-type theorem on self-dual bent functions. Des Codes Crypt 2012;63(3):295-304.

[181] Hyun JY, Lee H, Lee Y. Necessary conditions for the existence of regular p-ary bent functions. IEEE Trans Inform Theory 2014;60(3):1665-72.

[182] Ipatov VP, Kamaletdinov BZ. Groups of periodic bent function sequences. Izvestiya Vysshikh Uchebnykh Zavedenii Matematika 1998;3:26-32 [in Russian].

[183] Ipatov VP, Shebshayevich BV. GLONASS CDMA some proposals on signal formats for future GNSS air interface. In: InsideGNSS, July-August; 2010. p. 46-51.

[184] Ivanov AV. Usage of the reduced representation of Boolean functions in construction of their nonlinear approximations. Bull Tomsk State Univ Suppl 2007;23:31-5 [in Russian].

[185] Ivanov AV. Monomial approximations of plateaued functions. Prikladnaya Diskretnaya Matem (Appl Discret Math) 2008;1(1):10-4 [in Russian].

[186] Ivanov AV. The degree of proximity of the Boolean function reduced representation to the class of monomial functions according to basis selection. Prikladnaya Diskretnaya Matem (Appl Discret Math) 2009; (Suppl. 1):7-9 [in Russian].

[187] Ivanov AV. Noncloseness of hyper-bent-function class under the general linear group action. Matem Voprosy Kriptogr (Math Aspects Cryptogr) 2012;3(2):5-26 [in Russian].

[188] Iwata T, Kurosawa K. Probabilistic higher order differential attack and higher order bent functions. In: Proceedings of the 5th annual international conference on the theory and application of cryptology and information security (Singapore, Singapore, November 14-18, 1999). Lecture notes in computer science, vol. 1716. Berlin: Springer; 1999. p. 62-74.

[189] Jadda Z, Parraud P, Qarboua S. Quaternary cryptographic bent functions and their binary projection. Cryptogr Commun 2013;5(1):49-65.

[190] Jia W, Zeng X, Helleseth T, Li C. A class of binomial bent functions over the finite fields of odd characteristic. IEEE Trans Inform Theory 2012;58(9):6054-63.

[191] Kaliski B, Robshaw M. Linear cryptanalysis using multiple approximations. In: Advances in cryptology—CRYPTO'94, 14th annual international cryptology conference (Santa Barbara, California, USA, August 21-25, 1994). Lecture notes in computer science, vol. 839. Berlin: Springer; 1994. p. 26-39.

[192] Kasami T, Tokura N. On the weight structure of Reed-Muller codes. IEEE Trans Inform Theory 1970;16(6):752-9.

[193] Kantor WM. Codes, quadratic forms and finite geometries. Proc Symp Appl Math 1995;50:153-77. Available at: URL: http://darkwing.uoregon.edu/~kantor/.

[194] Kavut S, Maitra S, Yucel MD. Search for Boolean functions with excellent profiles in the rotation symmetric class. IEEE Trans Inform Theory 2007;53(5):1743-51.

[195] Kerdock AM. A class of low-rate non-linear binary codes. Inform Control 1972;20(2):182-7.

[196] Khoo K, Gong G, Stinson DR. A new family of gold-like sequences. In: Proceedings of ISIT—IEEE, international symposium on information theory (Lausanne, Switzerland, June 30-July 5, 2002); 2002, p. 181.

[197] Khoo K, Gong G, Stinson DR. A new characterization of semi-bent and bent functions on finite fields. Des Codes Crypt 2006;38(2):279-95.

[198] Kim S, Gil GM, No JS. New classes of bent functions and generalized bent functions. IEICE Trans Fundam Electron Commun Comput Sci 2004;E87.A(2): 480-8.

[199] Kim S, Gil GM, Kim KH, No J-S. Generalized bent functions constructed from partial spreads. In: Proceedings of the IEEE international symposium on information theory (Lausanne, Switzerland, June 30-July 5, 2002); 2002, p. 41.

[200] Kim S, Jung SY. Modified Reed-Muller coding scheme made from the bent function for dimmable visible light communications. IEEE Photon Technol Lett 2013; 25(1):11-3.

[201] Kim YS, Jang JW, No JS. On p-ary bent functions defined on finite fields. In: Proceedings of the conference on mathematical properties of sequences and other combinatorial structures (Los Angeles, USA, May 30-June 1, 2002). International series in engineering and computer science, vol. 726; 2003. p. 65-76.

[202] Knudsen L. Practically secure Feistel ciphers. In: Fast software encryption—FSE, the Cambridge security workshop (Cambridge, UK, December 9-11, 1993). Lecture notes in computer science, vol. 809. Berlin: Springer-Verlag; 1994. p. 211-21.

[203] Knudsen LR, Robshaw MJB. Non-linear approximation in linear cryptanalysis. In: Advances in cryptology—EUROCRYPT'96, workshop on the theory and application of cryptographic techniques (Saragossa, Spain. May 12-16, 1996). Lecture notes in computer science, vol. 1070. Berlin: Springer-Verlag; 1996. p. 224-36.

[204] Kolokotronis N, Limniotis K. Properties of bent functions Maiorana-McFarland functions with high second-order nonlinearity, Cryptology ePrint Archive, report 2011/212; 2011. Available at: URL: http://eprint.iacr.org/.

[205] Kolomeec NA. A graph of minimal distances of bent functions. In: Proceedings of CTCrypt-2015—current trends in cryptology (Kazan, Russia, June 03-05, 2015); 2015. 10 p.

[206] Kolomeec NA, Pavlov AV. Properties of bent functions on the minimal distance. Prikladnaya Diskretnaya Matem (Appl Discret Math) 2009;4:5-20 [in Russian].

[207] Kolomeec NA, Pavlov AV. Bent functions on the minimal distance. In: Proceedings of the IEEE region 8 international conference on computational technologies in electrical and electronics engineering (Irkutsk, Russia, July, 11-15, 2010); 2010. p. 145-9.

[208] Kolomeets NA. Enumeration of the bent functions on the minimal distance from the quadratic bent function. J Appl Ind Math 2012;6(3):306-17.

[209] Kolomeec NA. On a property of quadratic Boolean functions. Matem Voprosy Kriptogr (Math Aspects Cryptogr) 2014;5(2):79-85 [in English].

[210] Kolomeec NA. Constructions of bent functions on the minimal distance from the quadratic bent function. In: Proceedings of IEEE international symposium on information theory (St Petersburg, Russia, July 31-August 5, 2011); 2011. p. 647-51.

[211] Kolomeec NA. Upper bound for the number of bent functions on distance 2^k from an arbitrary bent function in $2k$ variables. Prikladnaya Diskretnaya Matem (Appl Discret Math) 2014;3:28-39 [in Russian].

[212] Kolomeec NA. Bent functions affine on subspaces and their metrical properties Ph.d. thesis; 2011 [in Russian].

[213] Kolomeec NA. Private communication; 2014.

[214] Korsakova EP. Graph classification for quadratic bent functions in 6 variables. Diskretn Anal Issled Oper (Discret Anal Oper Res) 2013;20(5):45-57 [in Russian]. English translation appears soon in J Appl Ind Math.

[215] Krotov DS. $\mathbb{Z}_4$-linear perfect codes. Diskretn Anal Issled Oper (Discret Anal Oper Res) 2000;7(4):78-90, [in Russian]. English translation is available at: URL: http://arxiv.org/abs/0710.0198.

[216] Krotov DS. $\mathbb{Z}_4$-linear Hadamard and extended perfect codes. In: Proceedings of the international workshop on coding and cryptography (Paris, France, January 8-12, 2001); 2001. p. 329-34.

[217] Ku-Cauich JC, Tapia-Recillas H. Systematic authentication codes based on a class of bent functions and the Gray map on a Galois ring. SIAM J Discret Math 2013;27(2):1159-70.

[218] Kumar PV, Scholtz RA, Welch LR. Generalized bent functions and their properties. J Combin Theory Ser A 1985;40(1):90-107.

[219] Kuznetsov YV, Shkarin RA. Reed-Muller codes (a survey of publications). Matem Voprosy Kibernetiki (Math Problems Cybern) 1995;6:5-50 [in Russian].

[220] Kuz'min AS, Markov VT, Nechaev AA, Shishkov AB. Approximation of Boolean functions by monomial functions. Discret Math Appl 2006;16(1):7-28.

[221] Kuz'min AS, Markov VT, Nechaev AA, Shishkin VA, Shishkov AB. Bent functions and hyper-bent functions over field with 2^l elements. Probl Inf Transm 2008; 44(1):12-33.

[222] Kuz'min AS, Nechaev AA, Shishkin VA. Bent and hyper-bent functions over the finite field. Trudy Diskret Matem (Bull Discret Math) M: Fizmatlit 2007;10:97-122 [in Russian].

[223] Kuz'min AS, Nechaev AA, Shishkin VA. Parameters of hyper-bent functions over field with 2^l elements. Trudy Diskret Matem (Bull Discret Math) M: Fizmatlit 2008;11: 47-59 [in Russian].

[224] Lahtonen J, McGuire G, Ward HN. Gold and Kasami-Welch functions, quadratic forms, and bent functions. Adv Math Commun 2007;1(2):243-50.

[225] Langevin P. Classification of Boolean quartics forms in eight variables; 2008. URL: http://langevin.univ-tln.fr/project/quartics/.

[226] Langevin P, Leander G. Monomial bent functions and Stickelberger's theorem. Finite Fields Appl 2008;14:727-42.

[227] Langevin P, Leander G, McGuire G. Kasami bent functions are not equivalent to their duals. Finite Fields Appl 2008;461:187-97. Proceedings of the 8th international conference on finite fields and applications (Melbourne, Australia, July 9-13, 2007).

[228] Langevin P, Leander G. Counting all bent functions in dimension eight 99270589265934370305785861242880. Des Codes Crypt 2011;59(1-3):193-205.

[229] Langevin P, Rabizzoni P, Véron P, Zanotti J. On the number of bent functions with 8 variables. In: Proceedings of the second international conference BFCA—Boolean functions: cryptography and applications (Rouen, France, March 13-15, 2006); 2006. p. 125-35.

[230] Leander NG. Monomial bent functions. IEEE Trans Inform Theory 2006;52(2): 738-43.

[231] Leander G, Kholosha A. Bent functions with Niho exponents. IEEE Trans Inform Theory 2006;52(12):5529-32.

[232] Leander NG, McGuire G. Construction of bent functions from near-bent functions. J Combin Theory Ser A 2009;116(4):960-70.

[233] Leveiller S, Zemor G, Guillot P, Boutros J. A new cryptanalytic attack for PN-generators filtered by a Boolean function. In: Proceedings of the selected areas of cryptography—SAC 2002. Lecture notes in computer science, vol. 2595; 2002. p. 232-49.

[234] Li N, Helleseth T, Tang X, Kholosha A. Several new classes of bent functions from Dillon exponents. IEEE Trans Inform Theory 2013;59(3):1818-31.

[235] Li N, Tang X, Helleseth T. New classes of generalized Boolean bent functions over $\mathbb{Z}_4$. In: Proceedings of the IEEE international symposium on information theory (Cambridge, USA, July 1-6, 2012); 2012. p. 841-5.

[236] Li N, Tang X, Helleseth T. New constructions of quadratic bent functions in polynomial form. IEEE Trans Inform Theory 2014;60(9):5760-7.

[237] Li Sh, Hu L, Zeng X. Constructions of p-ary quadratic bent functions. Acta Appl Math 2008;100(3):227-45.

[238] Li X, Hu Y, Gao J. Autocorrelation coefficients of two classes of semi-bent functions. Appl Math Inform Sci 2011;5(1):85-97.

[239] Lisonek P, Lu HY. Bent functions on partial spreads. Des Codes Crypt 2014; 73(1): 209-16.

[240] Liu F, Yue Q. A relationship between the nonexistence of generalized bent functions and class groups. Sci China Math 2010;53(1):213-22.

[241] Liu FM, Yue Q. The relationship between the nonexistence of generalized bent functions and diophantine equations. Acta Math Sin Engl Ser 2011;27(6):1173-86.

[242] Liu FM, Ma Z, Feng KQ. New results on non-existence of generalized bent functions (ii). Sci China Ser A: Math Phys Astron 2002;45(6):721-30.

[243] Logachev OA, Sal'nikov AA, Smyshlyaev SV, Yashchenko VV. Boolean functions in coding theory and cryptology. m.: Moscow center of uninterrupted mathematical education; 2012 [in Russian].

[244] Logachev OA, Sal'nikov AA, Yashchenko VV. Bent functions on a finite abelian group. Discret Math Appl 1997;7(6):547-64.

[245] Logachev OA, Sal'nikov AA, Yashchenko VV. Several characteristics of "nonlinearity" of group mappings. Diskretn Anal Issled Oper (Discret Anal Oper Res) 2001;8(1): 40-54 [in Russian].

[246] Logachev OA, Sal'nikov AA, Yashchenko VV. Cryptographic properties of discrete functions. In: Proceedings of the conference "Moscow University and development of cryptography in Russia", MSU, 2002. M.: MCUME; 2003. p. 174-99 [in Russian].

[247] Logachev OA, Sal'nikov AA, Yashchenko VV. Boolean functions in coding theory and cryptography. M.: Moscow center of uninterrupted mathematical education; 2004. Translated in English by AMS in the series "Translations of Mathematical Monographs" in 2012. ISBN-10: 0821846809, ISBN-13: 978-0821846803.

[248] Losev VV. Decoding of the bent function sequences by fast Hadamard transform. Radiotekhnika i Elektronika (Radio Eng Electron) 1987;32(7):1479-82 [in Russian].

[249] Ma W P LMHZFT. A new class of bent functions. IEICE Trans Fundam Electron Commun Comput Sci 2005;E88.A(7):2039-40.

[250] Maitra S, Sarkar P. Maximum nonlinearity of symmetric Boolean functions on odd number of variables. IEEE Trans Inform Theory 2002;48(9):2626-30.

[251] Maity S, Maitra S. Minimum distance between bent and 1-resilient Boolean functions. In: Proceedings of the 11th international workshop on fast software encryption (Delhi, India, February 5-7, 2004). Lecture notes in computer science, vol. 3017. Berlin: Springer; 2004. p. 143-60.

[252] Maity S, Maitra S. Minimum distance between bent and 1-resilient Boolean functions. Ars Combinatoria 2010;97:351-75.

[253] Mansoori SD, Bizaki HK. On the vulnerability of simplified AES algorithm against linear cryptanalysis. Int J Comput Sci Netw Secur 2007;7(7):257-63.

[254] Matsufuji S, Imamura K. Real-valued bent functions and its application to the design of balanced quadriphase sequences with optimal correlation properties. In: International symposium on applied algebra, Algebraic algorithms and error-correcting codes—AAECC-8 (Tokyo, Japan, August 20-24, 1990). Lecture notes in computer science, vol. 508; 1990. p. 106-12.

[255] Matsufuji S, Imamura K. Balanced quadriphase sequences with optimal periodic correlation properties constructed by real-valued bent functions. IEEE Trans Inform Theory 1993;39(1):305-10.

[256] Matsufuji S, Suehiro N. Symmetrical factorization of bent function type complex Hadamard matrices. IEICE Trans Fundam Electron Commun Comput Sci 1999;E82.A(12):2765-70.

[257] Matsui M. Linear cryptanalysis method for DES cipher. In: Advances in cryptology—EUROCRYPT'93, workshop on the theory and application of cryptographic techniques (Lofthus, Norway, May 23-27, 1993). Lecture notes in computer science, vol. 765. Berlin: Springer; 1994. p. 386-97.

[258] Matsui M. New structure of block ciphers with provable security against differential and linear cryptanalysis. In: Advances in cryptology—EUROCRYPT'96, workshop on the theory and application of cryptographic techniques (Saragossa, Spain, May 12-16, 1996). Lecture notes in computer science, vol. 1070. Berlin: Springer-Verlag; 1996. p. 205-18.

[259] Matsui M, Yamagishi A. A new method for known plaintext attack of FEAL cipher. In: Advances in cryptology—EUROCRYPT'92, workshop on the theory and application of cryptographic techniques (Balatonfured, Hungary, May 24-28, 1992). Lecture notes in computer science, vol. 658. Berlin: Springer-Verlag; 1993. p. 81-91.

[260] McFarland RL. A family of difference sets in non-cyclic groups. J Combin Theory Ser A 1973;15(1):1-10.

[261] MacWilliams FJ, Sloane NJA. The theory of error-correcting codes. Amsterdam, The Netherlands: North Holland; 1977. 762p.

[262] Meier W, Staffelbach O. Nonlinearity criteria for cryptographic functions. In: Advances in cryptography—EUROCRYPT'89. Lecture notes in computer science, vol. 434. Berlin: Springer; 1990. p. 549-62.

[263] Meng Q, Chen L, Fu FW. On homogeneous rotation symmetric bent functions. Discret Appl Math 2010;158(10):1111-7.

[264] Meng Q, Yang M, Zhang H, Cui JS. A novel algorithm enumerating bent functions. Discret Math 2008;308(23):5576-84. The first version is on Cryptology ePrint Archive, report 2004/274, Available at URL: http://eprint.iacr.org/.

[265] Meng Q, Zhang H, Wang Z. Designing bent functions using evolving computing. Acta Electron Sin 2004;11:1901-3.

[266] Meng Q, Zhang H, Yang MC, Cui J. On the degree of homogeneous bent functions. Discret Appl Math 2007;155(5):665-9.

[267] Mesnager S. A new family of hyper-bent Boolean functions in polynomial form. In: Proceedings of the 12th IMA conference on cryptography and coding (Cirencester, England, December 15-17, 2009). Lecture notes in computer science, vol. 5921; 2009. p. 402-17.

[268] Mesnager S. Recent results on bent and hyper-bent functions and their link with some exponential sums. In: Proceedings of the IEEE information theory workshop (Dublin, Ireland, August 30-September 3, 2010); 2010.

[269] Mesnager S. Hyper-bent Boolean functions with multiple trace terms. In: Proceedings of the 3rd international workshop on arithmetic of finite fields (Istanbul, Turkey, June 27-30, 2010). Lecture notes in computer science, vol. 6087; 2010. p. 97-113.

[270] Mesnager S. A new class of bent and hyper-bent Boolean functions in polynomial forms. Des Codes Crypt 2011;59(1-3):265-79.

[271] Mesnager S. Bent and hyper-bent functions in polynomial form and their link with some exponential sums and Dickson polynomials. IEEE Trans Inform Theory 2011;57(9):5996-6009.

[272] Mesnager S. Contributions on Boolean functions for symmetric cryptography and error correcting codes. Habilitation thesis; 2012.

[273] Mesnager S. Semi-bent functions from oval polynomials. In: Proceedings of the 14th IMA international conference (Oxford, UK, December 17-19, 2013); 2013. p. 1-15.

[274] Mesnager S. Several new infinite families of bent functions and their duals. IEEE Trans Inform Theory 2014;60(7):4397-407.

[275] Mesnager S, Flori JP. Hyper-bent functions via Dillon-like exponents. In: Proceedings of the IEEE international symposium on information theory (Cambridge, USA, July 1-6, 2012); 2012. p. 836-40.

[276] Mihaljevic M, Gangopadhyay S, Paul G, Imai H. An algorithm for the internal state recovery of Grain-v1. In: Proceedings of CECC'2011—Central European conference on cryptography (Debrecen, Hungary, June 30-July 2, 2011); 2011. p. 7-20.

[277] Millan W, Clark A, Dawson E. An effective genetic algorithm for finding highly nonlinear Boolean functions. In: First international conference on information and communications security—ICICS'97 (Beijing, China, November 11-14, 1997). Lecture notes in computer science, vol. 1334. Berlin: Springer-Verlag; 1997. p. 149-58.

[278] Millan W, Clark A, Dawson E. Smart hill climbing finds better Boolean functions. In: Workshop on selected areas in cryptology; 1997. p. 50-63. Workshop record.

[279] Moraga C, Stankovic M, Stankovic RS, Stojkovic S. Contribution to the study of multiple-valued bent functions. In: Proceedings of the IEEE 43rd international symposium on multiple-valued logic (Toyama, Japan, May 22-24, 2013); 2013. p. 340-5.

[280] Moraga C, Stankovic M, Stankovic RS, Stojkovic S. Hyper-bent multiple-valued functions. In: Proceedings of the conference on computer aided systems theory—EUROCAST 2013. Lecture notes in computer science, vol. 8112; 2013. p. 250-7.

[281] Muratovic-Ribic A, Pasalic E, Bajric S. Vectorial bent functions from multiple terms trace functions. IEEE Trans Inform Theory 2014;60(2):1337-47.

[282] Nakahara J. A linear analysis of Blowfish and Khufu. In: Information security practice and experience—ISPEC 2007, third international conference (Hong Kong, China, May 7-9, 2007). Lecture notes in computer science, vol. 4464; 2007. p. 20-32.

[283] Nakahara J, Preneel B, Vandewalle J. Experimental non-linear cryptanalysis; 2003. COSIC Internal Report.

[284] Nechaev AA. Kerdock code in a cyclic form. Discr Math Appl 1991;1(4): 365-84.

[285] Neumann T. Bent functions; 2006. Student's diploma.

[286] NSUCRYPTO. Official web-cite of the International Student's Olympiad in Cryptography. URL: www.nsucrypto.nsu.ru.

[287] Nyberg K. Constructions of bent functions and difference sets. In: Lecture notes in computer science, vol. 473; 1991. p. 151-60.

[288] Nyberg K. Perfect nonlinear S-boxes. In: Advances in cryptology—EUROCRYPT'1991, international conference on the theory and application of cryptographic techniques (Brighton, UK, April 8-11, 1991). Lecture notes in computer science, vol. 547. Berlin: Springer; 1991. p. 378-86.

[289] Nyberg K. Differentially uniform mappings for cryptography. In: Advances in cryptology—EUROCRYPT'1993, international conference on the theory and application of cryptographic techniques (Lofthus, Norway. May 23-27, 1993). Lecture notes in computer science, vol. 765. Berlin: Springer; 1994. p. 55-64.

[290] Nyberg K. Linear approximation of block ciphers. In: Advances in cryptology—EUROCRYPT'94, workshop on the theory and application of cryptographic techniques (Perugia, Italy, May 9-12, 1994). Lecture notes in computer science, vol. 950. Berlin: Springer; 1995. p. 439-44.

[291] Nyberg K, Hermelin M. Multidimensional Walsh transform and a characterization of bent functions. In: Proceedings of IEEE information theory workshop information theory for wireless networks (Solstrand, Norway, July 1-6, 2007); 2007. p. 83-6.

[292] Oblaukhov AK, On metrical supplement of an arbitrary set of the Boolean cube. Diskretn Anal Issled Oper (Discrete Anal Oper Res) in press (in Russian). English translation appears soon in J Appl Ind Math 2015.

[293] Olejár D, Stanek M. On cryptographic properties of random Boolean functions. J Univers Comput Sci 1998;4(8):705-17.

[294] Olsen JD, Scholtz RA, Welch LR. Bent-function sequences. IEEE Trans Inform Theory 1982;28(6):858-64.

[295] Pankratova I. Boolean functions in cryptography. Tomsk: Tomsk State University; 2014. 88p. [in Russian].

[296] Parker MG. The constabent properties of Golay-Davis-Jedwab sequences. In: Proceedings of the IEEE international symposium on information theory—ISIT'2000 (Sorrento, Italy, June 25-30, 2000); 2000, p. 302.

[297] Parker MG, Pott A. On Boolean functions which are bent and negabent. In: Sequences, subsequences, and consequences—SSC 2007, international workshop (Los Angeles, CA, USA, May 31-June 2, 2007). Lecture notes in computer science, vol. 4893. Berlin: Springer; 2007. p. 9-23.

[298] Pasalic E, Zhang WG. On multiple output bent functions. Inf Process Lett 2012;112(21):811-5.

[299] Paterson KG. Sequences for OFDM and multi-code CDMA: two problems in algebraic coding theory. In: Sequences and their applications—SETA 2001, second international conference (Bergen, Norway, May 13-17, 2001). Berlin: Springer; 2002. p. 46-71.

[300] Paterson KG. On codes with low peak-to-average power ratio for multicode CDMA. IEEE Trans Inform Theory 2004;50(3):550-8.

[301] Pavlov AV. Bent functions at the minimal distance and algorithms of constructing linear codes for CDMA, Cryptology ePrint Archive, report 2010/259; 2010. Available at: URL: http://eprint.iacr.org/.

[302] Pieprzyk JP. On bent permutations. Finite Fields Coding Theory Adv Commun Comput 1993;141:173-81.

[303] Pieprzyk JP, Qu CX. Fast hashing and rotation-symmetric functions. J Univer Comput Sci 1999;5:20-31.

[304] Poinsot L. Multidimensional bent functions. GESTS Int Trans Comput Sci Eng 2005;18(1):185-95.

[305] Poinsot L. Non Abelian bent functions. Cryptogr Commun 2012;4(1):1-23.

[306] Poinsot L, Harari S. Generalized Boolean bent functions. In: Progress in cryptology—INDOCRYPT 2004 (Chennai (Madras), India, December 20-22, 2004). Lecture notes in computer science, vol. 3348. Berlin: Springer; 2004. p. 107-19.

[307] Potapov VN. Cardinality spectra of components of correlation immune functions, bent functions, perfect colorings, and codes. Probl Inf Transm 2012;48(1): 47-55.

[308] Pott A, Tan Y, Feng T, Ling S. Association schemes arising from bent functions. Des Codes Crypt 2011;59(1-3):319-31.

[309] Preneel B, Van Leekwijck W, Van Linden L, Govaerts R, Vandevalle J. Propogation characteristics of Boolean functions. In: Advances in cryptology—EUROCRYPT'1990, international conference on the theory and application of cryptographic techniques (Aarhus, Denmark, May 21-24, 1990). Lecture notes in computer science, vol. 473. Berlin: Springer; 1991. p. 161-73.

[310] Preneel B. Analysis and design of cryptographic hash functions. Ph.d. thesis. Leuven, Belgium; 1993.

[311] Qu L, Li C. Minimum distance between bent and resilient Boolean functions. In: Proceedings of the 2nd international workshop on coding and cryptology (Zhangjiajie, China, June 1-5, 2009), vol. 473; 2009. p. 219-32.

[312] Qu L, Fu S, Dai Q, Li C. When a Boolean function can be expressed as the sum of two bent functions, Cryptology ePrint Archive, report 2014/048; 2014. Available at: URL: http://eprint.iacr.org/.

[313] Qu C, Seberry J, Pieprzyk J. Homogeneous bent functions. Discret Appl Math 2000;102(1-2):133-9.

[314] Riera C, Parker MG. Generalised bent criteria for Boolean functions (i). IEEE Trans Inform Theory 2006;52(9):4142-59.

[315] Rodier F. Asymptotic nonlinearity of Boolean functions. Des Codes Crypt 2006;40(1):59-70. Preprint is available at: URL: http://iml.univ-mrs.fr/editions/preprint2003/files/RodierFoncBool.pdf.

[316] Rodier F. Private communication; 2008.

[317] Rothaus O. On bent functions; 1966. IDA CRD W.P. No. 169.

[318] Rothaus O. On bent functions. J Combin Theory Ser A 1976;20(3):300-5.

[319] Ryabov VG. Ph.D. Dissertation. Moscow; 1984 (in Russian).

[320] Sakurai K, Furuya S. Improving linear cryptanalysis of LOKI91 by probabilistic counting method. In: Fast software encryption—FSE'97, 4th international workshop

(Haifa, Israel, January 20-22, 1997). Lecture notes in computer science, vol. 1267. Berlin: Springer; 1997. p. 114-33.

[321] Savicky P. On the bent Boolean functions that are symmetric. Eur J Comb 1994;15(4):407-10.

[322] Savicky P. Bent functions and random Boolean formulas. Discret Math 1995;147(1-3):211-34.

[323] Schmidt KU. Quaternary constant-amplitude codes for multicode CDMA. In: Proceedings of the IEEE international symposium on information theory—ISIT'2007 (Nice, France, June 24-29, 2007); 2007. p. 2781-5. Available at: URL: http://arxiv.org/abs/cs.IT/0611162.

[324] Selçuk AA. On probability of success in linear and differential cryptanalysis. J Cryptol 2008;21(1):131-47.

[325] Sharma D, Gangopadhyay S. On Kasami bent function, Cryptology ePrint Archive, report 2008/426; 2008. Available at: URL: http://eprint.iacr.org/.

[326] Shorin VV, Jelezniakov VV, Gabidulin EM. Linear and differential cryptanalysis of Russian GOST. In: Proceedings of the international workshop on coding and cryptography (Paris, France, January 8-12, 2001); 2001. p. 467-76.

[327] Sidel'nikov VM. On the mutual correlation of sequences. Probl Cybern 1971;24:15-42 (in Russian).

[328] Sidelnikov VM. On extremal polynomials used in code size estimation. Probl Inf Transm 1980;16(3):174-86.

[329] Singh D, Bhaintwal M, Singh BK. Some results on q-ary bent functions. Int J Comput Math 2013;90(9):1761-73.

[330] Smidth DH, Hunt FH, Perkins S. Exploiting spatial separations in CDMA systems with correlation constrained sets of Hadamard matrices. IEEE Trans Inform Theory 2010;56(11):5757-61.

[331] Solé P, Tokareva N. Connections between quaternary and binary bent functions, Cryptology ePrint Archive, report 2009/544; 2009. Available at: URL: http://eprint.iacr.org/.

[332] Solodovnikov VI. Bent functions from a finite Abelian group into a finite Abelian group. Discret Math Appl 2002;12(2):111-26.

[333] Solodovnikov VI. On the coincidence of the class of bent-functions with the class of functions which are minimally close to linear functions. Prikladnaya Diskretnaya Matem (Appl Discret Math) 2012;3:25-33 [in Russian].

[334] Solodovnikov VI. On primary functions which are minimally close to linear functions. Matem Voprosy Kriptogr (Math Aspects Cryptogr) 2012;2(4):97-108 [in Russian].

[335] Stănică P. Chromos, Boolean functions and avalanche characteristics. Ph.d. thesis; 1998.

[336] Stănică P. On the nonexistence of homogeneous rotation symmetric bent Boolean functions of degree greater than two. In: Proceedings of NATO Adv. Stud. Instit.—Boolean functions (ASI07, Moscow, Russia); 2008.

[337] Stănică P, Gangopadhyay S, Chaturvedi A, Gangopadhyay AK, Maitra S. Nega-Hadamard transform bent and negabent functions. In: Proceedings of the 6th international conference on sequences and their applications (Paris, France, September 13-17, 2010). Lecture notes in computer science, vol. 6338; 2010. p. 359-72.

[338] Stănică P, Gangopadhyay S, Chaturvedi A, Gangopadhyay AK, Maitra S. Investigations on bent and negabent functions via the nega-Hadamard transform. IEEE Trans Inform Theory 2012;58(6):4064-72.

[339] Stănică P, Maitra S. A constructive count of rotation symmetric functions. Inform Process Lett 2003;88:299-304.

[340] Stănică P, Maitra S. Rotation symmetric Boolean functions—count and cryptographic properties. Discret Appl Math 2008;156:1567-80. Preliminary version appeared in Electron Notes Discret Math 2003;15:141-147.

[341] Stănică P, Maitra S, Clark J. Results on rotation symmetric bent and correlation immune Boolean functions. In: Fast software encryption (Delhi, 2004). Lecture notes in computer science, vol. 3017. Berlin: Springer; 2004. p. 161-77.

[342] Stănică P, Martinsen T, Gangopadhyay S, Singh BK. Bent and generalized bent Boolean functions. Des Codes Cryptogr 2013;69:77-94.

[343] Stankovic S, Stankovic M, Astola J. Representation of multiple-valued bent functions using Vilenkin-Chrestenson decision diagrams. In: Proceedings of the 41st IEEE international symposium on multiple-valued logic (Tuusula, Finland, May 23-25, 2011); 2011. p. 62-8.

[344] Stoyanov BP, Kordov KM. Cryptanalysis of a modified encryption scheme based on bent Boolean function and feedback with carry shift register. In: Proceedings of the 5th international conference for promoting the application of mathematics in technical and natural sciences (Varna, Bulgaria, June 24-29, 2013), vol. 1561; 2013. p. 373-7.

[345] Su W, Pott A, Tang X. Characterization of negabent functions and construction of bent-negabent functions with maximum algebraic degree. IEEE Trans Inform Theory 2013;59(6):3387-95.

[346] Sun G, Wu C. Construction of semi-bent Boolean functions in even number of variables. Chin J Electron 2009;18(2):231-7.

[347] Sun G, Wu C. Comments on "Monomial Bent Functions". IEEE Trans Inform Theory 2011;57(6):4014-5.

[348] Tan Y, Pott A, Feng T. Strongly regular graphs associated with ternary bent functions. J Combin Theory Ser A 2010;117(6):668-82.

[349] Tang C, Lou Y, Qi Y, Xu M, Guo B. A note on semi-bent and hyper-bent Boolean functions. In: Proceedings of the 9th China international conference on information security and cryptology (Guangzhou, China, November 27-30, 2013). Lecture notes in computer science, vol. 8567. Berlin: Springer; 2013. p. 3-21.

[350] Tang D, Carlet C, Tang X. On the second-order nonlinearities of some bent functions. Inform Sci 2013;223:322-30.

[351] Tarannikov YV. Combinatorial properties of discrete structures and applications to cryptology. M.: Moscow center of uninterrupted mathematical education; 2011 [in Russian].

[352] Teng JH, Li SQ, Huang XY. The kth-order quasi-generalized bent functions over ring Z(p). In: Proceedings of the 1st SKLOIS conference on information security and cryptology (Beijing, China, December 15-17, 2005). Lecture notes in computer science, vol. 3822. Berlin: Springer; 2005. p. 189-201.

[353] Teng JH, Huang XY, Li XR. Spectrum distribution characteristics of kth-order generalized quasi-bent functions over ring Z(p). In: Proceedings of the 14th youth conference on communication 2009 (Dalian, China, July 24-26, 2009); 2009. p. 425-8.

[354] Tokareva NN. Bent functions with more strong nonlinear properties: k-bent functions. J Appl Ind Math 2008;2(4):566-84.

[355] Tokareva NN. k-Bent functions: from coding theory to cryptology. In: Proceedings of the 2008 IEEE region 8 international conference on computational technologies in electrical and electronics engineering (Novosibirsk, Russia, July 21-25, 2008); 2008. p. 36-40.

[356] Tokareva NN. k-Bent functions and quadratic approximations in block ciphers. In: Proceedings of the fourth international conference BFCA—Boolean functions: cryptography and applications (Copenhagen, Denmark, May 19-21, 2008); 2008. p. 132-48.

[357] Tokareva NN. On quadratic approximations in block ciphers. Probl Inf Transm 2008;44(3):266-86.

[358] Tokareva NN. Description of k-bent functions in four variables. J Appl Ind Math 2009;3(2):284-9.

[359] Tokareva NN. Bent functions: results and applications. a survey of publications. Prikladnaya Diskretnaya Matem (Appl Discret Math) 2009;2(1):15-37 [in Russian].

[360] Tokareva NN. The group of automorphisms of the set of bent functions. Discret Math Appl 2010;20(5-6):655-64.

[361] Tokareva NN. Generalizations of bent functions. A survey. J Appl Ind Math 2011;5(1):110-29. Available also on Cryptology ePrint Archive, report 2011/111.

[362] Tokareva NN. Nonlinear Boolean functions: bent functions and their generalizations. Saarbrucken, Germany: LAP LAMBERT Academic Publishing; 2011. 170p. [in Russian].

[363] Tokareva NN. On the number of bent functions from iterative constructions: lower bounds and hypotheses. Adv Math Commun 2011;5(4):609-21.

[364] Tokareva NN. Duality between bent functions and affine functions. Discret Math 2012;312:666-70.

[365] Tokareva NN. On decomposition of a Boolean function into sum of bent functions. Sib Electron Math Rep 2014;11:745-51.

[366] Tokareva NN. On decomposition of the dual bent function into sum of two bent functions. Prikladnaya Diskretnaya Matem (Appl Discret Math) 2014;4:59-61 [in Russian].

[367] Tu Z, Deng Y. A conjecture about binary strings and its applications on constructing Boolean functions with optimal algebraic immunity. Des Codes Crypt 2011;60(1): 1-14.

[368] Tuzhilin ME. Almost perfect nonlinear functions. Prikladnaya Diskretnaya Matem (Appl Discret Math) 2009;3:14-20 [in Russian].

[369] Wada T. Characteristic of bit sequences applicable to constant amplitude orthogonal multicode systems. IEICE Trans Fundam Electron Commun Comput Sci 2000;E83-A(11):2160-4.

[370] Wang B, Tang C, Qi Y, Yang Y, Xu M. A new class of hyper-bent Boolean functions with multiple trace forms, Cryptology ePrint Archive, report 2011/600; 2011. Available at: URL: http://eprint.iacr.org/.

[371] Wang L, Zhang J. A best possible computable upper bound on bent functions. J West China 2004;33(2):113-5 [in Chinese].

[372] Wang Q, Tan CH. A note on the algebraic immunity of the Maiorana-McFarland class of bent functions. Inf Process Lett 2012;112(22):869-71.

[373] Wang X, Zhou J, Zang Y. A note on homogeneous bent functions. In: Proceedings of the 8th ACIS international conference on software engineering, artificial intelligence, networking and parallel/distributed computing (Qungdao, China, July 30—August 1, 2007); 2007. p. 138-42.

[374] Wang X, Zhou J. Generalized partially bent functions. In: Future generation communication and networking (Jeju-Island, Korea, December 6-8, 2007); 2007. p. 16-21.

[375] Wang Z, Xu C. Constructing bent functions by GH-matrices. In: Proceedings of the 3rd international workshop on matrix analysis and applications (Hangzhou, China, July 9-13, 2009); 2009. p. 9-12.

[376] Weng G, Feng R, Qiu W, Zheng Z. The ranks of Maiorana-McFarland bent functions. Sci China Ser A: Math 2008;51(9):1726-31.

[377] Weng G, Feng R, Qiu W. On the ranks of bent functions. Finite Fields Appl 2007;13(4):1096-116.

[378] Wolfmann J. $\mathbb{Z}_4$-version of the binary Maiorana-McFarland bent functions. In: Proceedings of the IEEE international symposium on information theory (Cambridge, USA, August 16-22, 1998); 1998, p. 401.

[379] Wolfmann J. Bent functions and coding theory. In: Proceedings of the NATO advanced study institute on difference sets sequences and their correlation properties (Bad Winsheim, Germany, August 2–14, 1998), vol. 542; 1999. p. 393-418.

[380] Wolfmann J. Cyclic codes aspects of bent functions. Finite Fields Theory Appl 2010;518:363-84.

[381] Wolfmann J. Special bent and near bent functions. Adv Math Commun 2014; 8(1):21-33.

[382] Xia T, Seberry J, Pieprzyk J, Charnes C. Homogeneous bent functions of degree n in $2n$ variables do not exist for $n > 3$. Discret Appl Math 2004;142(1-3):127-32.

[383] Xia Y, Sui Y, Hu J. A generalization of the bent-function sequence construction. In: Proceedings of the 6th international symposium on neural networks (Wuhan, China, May 26-29, 2009). Lecture notes in computer science, vol. 5553; 2009. p. 557-66.

[384] Xiang Q. Maximally nonlinear functions and bent functions. Des Codes Crypt 1999;17(1-3):211-8.

[385] Yang M, Meng Q, Zhang H. Evolutionary design of trace form bent functions, Cryptology ePrint Archive, report 2005/322; 2005. Available at: URL: http://eprint. iacr.org/.

[386] Yashchenko VV. On the propagation criterion for Boolean functions and on bent functions. Probl Inf Transm 1997;33(1):62-71.

[387] Yashchenko VV. On two characteristics of nonlinearity of Boolean mappings. Diskretn Anal Issled Oper (Discret Anal Oper Res) 1998;5(2):90-6 [in Russian].

[388] Youssef AM. Generalized hyper-bent functions over $GF(p)$. Discret Appl Math 2007;155(8):1066-70.

[389] Youssef A, Gong G. Hyper-bent functions. In: Advances in cryptology—EUROCRYPT'2001, international conference on the theory and application of cryptographic techniques (Innsbruk, Austria, May 6-10, 2001). Lecture notes in computer science, vol. 2045. Berlin: Springer; 2001. p. 406-19.

[390] Yu NY, Gong G. Constructions of quadratic bent functions in polynomial forms. IEEE Trans Inform Theory 2006;52(7):3291-9.

[391] Yu NY, Gong G. Quadratic bent functions of polynomial forms and their applications to bent sequences. In: Proceedings of the 23rd biennial symposium on communications (Kingston, Canada, May 29-June 1, 2006; 2006. p. 128-31.

[392] Zhang B, Lü S. I/O correlation properties of bent functions. Sci China Ser E: Technol Sci 2000;43(3):282-6.

[393] Zhang W, Lv C, Xiao G. Nearly bent functions. Int J Comput Math 2011;88(5):943-9.

[394] Zhang X, Guo H, Gao Z. Characterizations of bent and almost bent function on $Z(p)(2)$. Graphs Combinator 2011;27(4):603-20.

[395] Zhang Y, Liu M, Lin D. On the nonexistence of bent functions. Int J Found Comput Sci 2011;22(6):1431-8.

[396] Zhang Z. The correlation between the internal construction and the property of bent functions in cryptographic system. In: Proceedings of the 2nd IEEE international conference on computer science and information technology (Beijing, China, August 8-11, 2009); 2009. p. 329-32.

[397] Zhao Y. Some properties of bent functions. Acta Math Sin Engl Ser 2007;23(3):389-94.

[398] Zhao Y, Li H. On bent functions with some symmetric properties. Discret Appl Math 2006;154(17):2537-43.

[399] Zheng D, Yu L, Hu L. On a class of binomial bent functions over the finite fields of odd characteristic. Appl Algebra Eng Commun Comput 2013;24(6):461-75.

[400] Zheng D, Yu L, Hu L. Quadratic bent and semi-bent functions over finite fields of odd characteristic. Chin J Electron 2014;23(4):767-72.

[401] Zheng D, Zeng X, Hu L. A family of p-ary binomial bent functions. IEICE Trans Fundam Electron Commun Comput Sci 2011;E94.A(9):1868-72.

[402] Zheng Y, Pieprzyk J, Seberry J. Haval—a one-way hashing algorithm with variable length of output (extended abstract). In: Proceedings of "Advances in cryptology—AUSCRYPT'92,". Lecture notes in computer science, vol. 718; 1993. p. 83-104.

[403] Zheng Y, Zhang XM. Relationships between bent functions and complementary plateaued functions. In: ICISC'99—international conference on information security and cryptology (Seoul, Korea, December 9-10, 1999). Lecture notes in computer science, vol. 1787. Berlin: Springer; 2000. p. 60-75.

[404] Zheng Y, Zhang XM. On plateaued functions. IEEE Trans Inform Theory 2001;47(3):1215-23.

[405] Zhou J, Mow WH, Dai X. Bent functions and codes with low peak-to-average power ratio for multi-code CDMA. In: Proceedings of the 17th international symposium on applied algebra, Algebrais algorithms and error-correcting codes (Bangalore, India, December 16-20, 2007). Lecture notes in computer science, vol. 4851. Berlin: Springer; 2007. p. 60-71.

INDEX

Printed in the United States
By Bookmasters